Emerging Trends in Mechatronics

Series Editor

Aydin Azizi, Oxford, UK

This book series helps understand how Mechatronics will impact the practice and research of developing advanced techniques to model, control and optimize complex systems. It presents recent state of advances in mechatronics and related technologies including: Automatic Control, Robotics, Agent Based systems, Smart Manufacturing and Industry 4.0. The selected topics give an overview of the state-of-the-art and present new research results and prospects of the future development in the interdisciplinary field of mechatronic systems. The book series will provide up-to-date and useful knowledge for researchers and engineers involved in mechatronics and related fields.

Within the above dimensions, the scope of the book series welcomes high quality original research and review articles that cover a broad range of topics related to theoretical and applied aspects of modern mechatronics. Potential topics include, but are not limited to the following:

Controller design

Many mechatronic systems have challenging control difficulties due to the high nonlinear structures and time-varying dynamic behaviors. In addition to these, there are external disturbances, which cannot be predicted and change according to uncontrolled working environments. Conventional controllers are insufficient to solve the problems and to compensate the environmental disturbances, therefore designing efficient controllers for complex systems has a great importance in the current book series.

Sensor Design & Data Collection Approaches

Essential part of a mechatronics system is the measurement system that senses the variations in the physical parameters, such as temperature, pressure, displacement, and so on, and converts it to voltage or current. In mechatronic systems, accurate measurement of the dynamic variables plays a vital role for the actuators to function properly, therefore on of the aims of this book series, to focus on understanding what should be done to design accurate sensors and efficient data collection methods to improve the performance of mechatronic systems.

Optimization Techniques

Designing a complex mechatronic system involves multiple variables, objectives, constraints, and evaluation criteria as well as their nonlinearly coupled relationships. One of the challenging research areas of this book series is utilizing optimization techniques to design an optimal mechatronic system.

Artificial Intelligence

From the point of view of dynamics, mechatronic systems can be characterized by model uncertainties, high nonlinearities, complicated coupling, and stringent performance requirements, among others. To deal with these facts, this book series would like to focus on investigating and utilizing Artificial Intelligence techniques such as Neural Networks, Fuzzy logic, Neuro-fuzzy, evolutionary techniques etc. to model, design and optimize complex mechatronic systems.

Robotics

In the field of Robotics, there is a need of focusing on kinematic, dynamic, control and optimization of innovative robotic systems. This book series can introduce new and novel robot designs such as hybrid (serial-parallel) redundant manipulators that have a high potential of application in special fields like space robots. State of the art researches on selection of optimum kinematic solutions and path planning based on such an optimization, dynamic analysis and new control methods developments and application of artificial intelligence in this field can be published in this book series.

Smart manufacturing

Application of networked sensors and machinery equipment using internet and intranet in order to monitor and automation of the manufacturing process and improve the performance efficiency and is a point of interest. Advanced robotics, multi input-multi output control approaches combined with the artificial intelligence and optimization techniques, big data processing approaches, multi-scale modelling and simulation of dynamic sub-systems etc. are the hot topics that will be focused in this book series.

AGVs

Automated Guided Vehicle (AGVs) are going to be very popular is industry logistics, surveillance, defense and social applications in order to make processes safer and more cost efficient. Innovative AGVs designs for various missions, Robot Operating System usage on AGV control, Indoor and outdoor navigation and control algorithms, intelligent path planning techniques, sensor fusion, simultaneous application of GPS, INS and offline maps are the topics that are planned to be published in this book series.

Masoomeh Mirrashid
Danial Jahed Armaghani • Aydin Azizi
Editors

Optimization and Adaptive Control Algorithms

Springer

Editors
Masoomeh Mirrashid
College of Engineering
Abu Dhabi University
Al Ain, United Arab Emirates

Danial Jahed Armaghani
School of Civil and Environmental Engineering
University of Technology Sydney
Sydney, NSW, Australia

Aydin Azizi
School of Engineering, Computing and Mathematics
Oxford Brookes University
Oxford, UK

ISSN 2731-4855 ISSN 2731-4863 (electronic)
Emerging Trends in Mechatronics
ISBN 978-981-95-8386-7 ISBN 978-981-95-8387-4 (eBook)
https://doi.org/10.1007/978-981-95-8387-4

This Springer imprint is published by the registered company Springer Nature Singapore Pte Ltd.
The registered company address is: 152 Beach Road, #21-01/04 Gateway East, Singapore 189721, Singapore

Contents

Fourier-Based Modeling of Norway's Fishing Load Capacity: Nonlinear and Cyclical Dynamics

Irina Georgescu and Jani Kinnunen

Abstract This study combines the Fourier-augmented Kernel-based regularized least squares (Fourier-KRLS) with the Fourier autoregressive distributed lag (FARDL) to model the fishing load capacity factor (FLCF) in Norway.

Traditional linear models often fail to model short-run oscillations and long-run dependencies, as marine resource pressures tend to be nonlinear and cyclical.

Our approach corrects this drawback by combining the spectral filtering capability of the Fourier transform with the KRLS model's resilience to nonlinearities. We discuss the influence of urbanization (URB), foreign direct investment (FDI), and GDP per capita on Norway's FLCF for 1990–2024. The FARDL model confirms cointegration among FLCF and determinants. URB has a positive long-run effect on FLCF (25.94), while GDP reduces FLCF (−6.92) due to environmental controls and infrastructure. FDI has a weak positive effect (0.11). The speed of adjustment to long-run equilibrium is quick, and the Fourier term values are significant. The Fourier-KRLS output shows that URB has the maximum marginal impact on FLCF. The Fourier terms, statistically significant, show hidden cyclical patterns. The Fourier-KRLS model outperforms the standard KRLS. Our findings prove the applicability of nonlinearity and cycles in modeling environmental sustainability.

Keywords Kernel-based regularized least squares · Fourier approximation Ecological footprint · Urbanization · Average marginal effects

---

I. Georgescu (✉)
Bucharest University of Economics, Bucharest, Romania
e-mail: irina.georgescu@csie.ase.ro

J. Kinnunen
LUT Business School, Lappeenranta, Finland
e-mail: jani.kinnunen@abo.fi

M. Mirrashid et al. (eds.), *Optimization and Adaptive Control Algorithms*, Emerging Trends in Mechatronics, https://doi.org/10.1007/978-981-95-8387-4_1

1 Introduction

Global sustainability research continues to investigate the causes that contribute to pollution in the environment of ocean and coastal ecosystems. Enhanced global warming, glacial retreat, reduced biodiversity, and increased flood and fire frequency are some of the environmental degradations occurring in recent decades [1]. Urbanization, industrialization, and utilization of nonrenewable resources caused the above environmental degradations.

Most of the research has been centered on the environmental indicator represented by CO_2 emissions due to their impact on global warming and climatic change. Environmental degradation goes beyond air pollution, affecting ecosystems from human activities. Wackernagel and Rees [2] introduced the ecological footprint (EF) as an indicator that measures human activities' impact on the environment. EF comprises the amount of consumed natural resources and the area of biologically productive land to provide this consumption.

As per the study [3], EF contains six indicators: grazing land, cropland, fishing grounds, forest area, built-up land, and carbon footprint. There are several studies on the determinants of EF ([4, 5], etc.). EF consists of two components: the demand side, quantified by the ecological footprint, and the supply side, which is quantified by its biocapacity [6]. Siche et al. [7] introduced the concept of load capacity factor (LCF) as an improved environmental measure. LCF identifies a country or region's ability to sustain its population at current living conditions. LCF is expressed as a ratio between biocapacity and EF.

A value of LCF less than 1 means an unsustainable environmental scenario, and a value of LCF above 1 means the system stays within sustainable limits [7, 8]. Research on LCF has been done regarding other Nordic countries as well. Kinnunen et al. [9] used the ARDL technique and frequency domain analysis to discuss how LCF in the electric grid evolves in Finland. Their research detected the nonlinear behavior of GDP growth, nuclear energy, and innovation of the system's efficiency. The fishing ground footprint shows the condition of the environmental ecosystem, representing the extent of the water area needed in order to fulfill the marine species demand [10].

FF has been the focus of research in recent years due to increasing concerns with overfishing, biodiversity loss, and climate change [11]. In high-income nations such as Norway, where marine resources are not only economically valuable but also ecologically sensitive, one has to make an estimation of the influence of economic variables on fishing-led degradation quantified by the load fishing capacity factor (FLCF).

Figure 1 illustrates the trend of Norway's FLCF during 1990–2024. This trend is increasing, particularly after 2005, and its peak is reached after 2020. This means that fishing effort pressure has been better aligned with the sustainable capacity of the ecosystem. The lower FLCF values reported in the 1990s and early 2000s emphasize the early stages of sustainable activity or lower exploitation intensity. The trend increase has been attributed to higher demand or to a limited capacity of

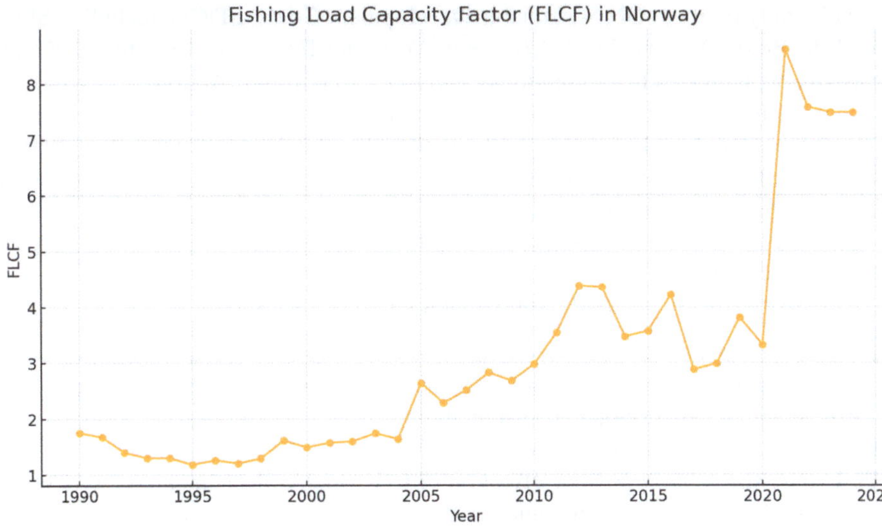

Fig. 1 Fishing load capacity factor in Norway. (Source: Own computation)

recovery of marine ecosystems. The trend has become a major issue in recent decades.

Sarkodie and Owusu [12] remark important pressures on global fish systems. From 2021, the top ten fish-producing countries, including Norway, China, Indonesia, India, Vietnam, Peru, the USA, Russia, Japan, and Chile, accounted for 68.88% of global fish production. Norway is one of the main contributors in global marine exports. Global fish consumption increased by 82.29% from 98,784 to 180,075 thousand tonnes between 1990 and 2021. This growth is mainly due to developing countries (a 163.14% increase), while including Norway had a 17.3% decline in fish consumption during 1990–2021.

Traditional parametric models try to capture the complexity of human--environment systems. They assume linearity and steady effects, which are not usually supported by empirical data [13].

In contrast, machine learning techniques, namely nonparametric techniques, generalize more by allowing for nonlinear interaction and heterogeneous effects. Kernel-based regularized least squares (KRLS) yields marginal effect estimates without functional form specification. KRLS has been applied more frequently in empirical social science [14].

Seasonal activity, gradual policy changes, or international treaties are the basis for the smooth, unobservable time cycles of environmental variables [15]. To capture such hidden behavior, we extend KRLS by adding Fourier terms, which capture long-term, periodical trends using trigonometric expansions. The new Fourier--KRLS model allows for a more flexible nonlinear estimation and representation of gradual structural change.

This study refers to UN Sustainable Development Goals (SDGs), including SDG 14 (Life Below Water), SDG 8 (Decent Work and Economic Growth), SDG 11 (Sustainable Cities and Communities), and SDG 13 (Climate Action).

In this study, we apply Fourier-based methods to investigate how GDP, URB, and FDI impact Norway's FLCF. The findings show that GDP has a lower effect, which is sensitive to cyclical adjustments. URB is associated with lower fishing pressure. FDI displays no significant impact.

More accurate estimates are obtained by the Fourier-KRLS model, which reflects residual fluctuations.

This work aims to verify the following three hypotheses:

H1: GDP per capita has a negative effect on FLCF, suggesting that economic development contributes to reducing pressure on marine ecosystems by improving efficiency and environmental sustainability.
H2: FDI has a positive influence on FLCF.
H3: Urbanization has a positive effect on FLCF, which reflects an increase in pressure on marine ecosystems and, consequently, a detrimental impact on their sustainability.

The chapter contributes to maritime environmental studies by filling in several research gaps. To evaluate the health of marine ecosystems represented by FLCF, we introduce a novel method, namely, the combination of FARDL and Fourier--KRLS. FLCF includes the demand side (FF) and the supply side (marine biocapacity).

This chapter is organized as follows: Sect. 2 gives an overview of recent research into environmental degradation with a focus on fishing footprint and sustainability indicators. Section 3 outlines the methodological framework by presenting both the FARDL model and the KRLS approach, as well as its extension through the Fourier methodology. Section 4 presents the empirical evidence from Norway for 1990–2024, including model estimation, diagnostic testing, and marginal effect discussion in a comparative context. Section 5 ends the chapter with conclusion, policy recommendations, and future research directions.

2 Literature Review

The investigation of environmental degradation has been developed in recent years using novel indicators and country case studies. Yilanci et al. [16] explored the pollution haven and halo hypotheses for Indonesia using the fishing footprint (FF) as an environmental indicator. Their findings confirmed the EKC and proved a positive impact of FDI in reducing environmental degradation in the fishery sector through technology transfer and more efficient management practices. Trade openness in fisheries diminishes FF, proving that trade expansion can encourage sustainable development. Wirajing et al. [17] debate the effectiveness of environmental taxation in reducing FF in 23 African countries by means of a smoothed instrumental-variable quantile regression approach. The study concludes that environmental taxation is effective in reducing ecological pressure only until the 0.75th quantile, especially in the case of middle-income and coastal countries. Its effect is limited in low-income and landlocked countries.

Mance et al. [18] prove the existence of an EKC for fish stock condition, showing that fish stocks worsen at earlier levels of human development but improve as development continues past a turning point. Their dynamic panel GMM estimation across 32 economies also reveals the beneficial effect of FDI, education, and R&D spending on fishery sustainability. Mustafa et al. [19] analyze the environmental sustainability of Indonesia's fishery industry in terms of the effect of fishing production, population growth, and container port traffic on the FLCF of Indonesia from 2000 to 2022. The study uses the advanced fractional frequency Fourier augmented autoregressive distributed lag model (FFF-ARDL), which results in an inverted U-shaped relationship between fishing production and FLCF, with more production in the initial stages being directed to the sustainability threshold. When the turning point exceeds 7.87 million metric tons, the environment gets destabilized. Other results of the study are that an increase of 1% in port activity and population decreases FLCF by 0.12% and 0.41%, respectively.

Yilanci et al. [10] conduct a sectoral analysis and examine the EKC in China. The analysis confirms the existence of the EKC between economic growth and FF. The study lays emphasis on the importance of subcomponents of EF like FF to have a better understanding of resource use in sectors such as fisheries.

Georgescu et al. [20] validated the existence of the EKC hypothesis for Norway, taking into account EF as a comprehensive sustainability indicator. Canbay et al. [21] investigated the causal link between education, measured by the net schooling rate, and aquaculture production in the five Turkish coastal provinces from 2009 to 2022. Using the panel bootstrap Granger causality test to account for structural heterogeneity and cross-sectional dependence, their empirical study found an asymmetric regionally varying effect of education on aquaculture production. A result of the study was that in the Aegean and Mediterranean regions, higher education contributes to fisheries and aquaculture production. In other regions such as the Western Black Sea, schooling was associated with a decline in production.

Obekpa et al. [22] use the KRLS model to assess the environmental cost of food security in Nigeria. They conclude that population growth and food production increase cropland and forest land. Economic development corresponds to higher carbon and FF.

Solarin et al. [23] use the fractional integration model for a set of 25 OECD countries from 1961 to 2017 and study whether shocks to fishery product consumption are permanent or temporary. Their results exhibit an increasing fishery product consumption. Most countries present evidence of mean reversion; that is, short-term policies are more effective than long-term ones in influencing the trajectory of fishery consumption.

The main conclusion of some of these studies is that Fourier-based models are efficient when dealing with heterogeneity and nonlinearity of environmental indicators. Also, Fourier-based methods are more flexible than standard regression, particularly for multiple thresholds or regime shifts [24]. Sustainable development in various national contexts is enriched by combining sector-specific ecological indicators like FF and FLCF with Fourier-based econometric methods such as FARDL or Fourier-KRLS.

3 Methodology

3.1 Input Data

3.1.1 Fourier ARDL

The econometric model used in this study is

$$FLCF_t = a_0 + a_1 GDP_t + a_2 FDI_t + a_3 URB_t + \varepsilon_t, \tag{1}$$

where ε_t is the error term.

Throughout the chapter, we will use logarithmic transformation of the variables. The log transformation contributes to stabilizing the variance, mitigates the influence of outliers, and reduces skewness. Coefficients are interpreted as elasticities [25].

The ARDL model by Pesaran et al. [26] is

$$\begin{aligned}\Delta FLCF_t = a_0 + \sum_{k=1}^{n} a_{1k}\Delta FLCF_{t-k} + \sum_{k=0}^{p} a_{2k}\Delta GDP_{t-k} + \sum_{k=0}^{q} a_{3k}\Delta FDI_{t-k} + \sum_{k=0}^{r} a_{4k}\Delta URB_{t-k} + \\ +\lambda_1 FLCF_{t-1} + \lambda_2 GDP_{t-1} + \lambda_3 FDI_{t-1} + \lambda_4 URB_{t-1} + \varepsilon_t\end{aligned} \tag{2}$$

In Eq. (2), Δ represents the first difference operator. The lag lengths n, p, q, and r will be determined later based on the Akaike Information Criterion (AIC).

λ_i, $i = 1, \ldots, 4$ are long-term coefficients.

Yilanci et al. [27] suggested the inclusion of a Fourier function to capture structural breaks. The Fourier approach is a flexible tool for handling smooth and unknown breaks in data:

$$d(t) = \sum_{k=1}^{n} ak \sin\frac{2\pi kt}{T} + \sum_{k=1}^{n} bk \cos\frac{2\pi kt}{T} \tag{3}$$

In Eq. (3), t is the trend, T is the total number of periods, n is the total number of frequencies, and k is the selected number of frequencies. Ludlow and Enders [28] recommend a single frequency for the analysis:

$$d(t) = \gamma_1 \sin\frac{2\pi kt}{T} + \gamma_2 \cos\frac{2\pi kt}{T} \tag{4}$$

In this work, the FARDL model by Enders and Lee [29] has the equation:

$$\begin{aligned}\Delta FLCF_t = a_0 + \gamma_1 \sin\frac{2\pi kt}{T} + \gamma_2 \cos\frac{2\pi kt}{T} + \sum_{k=1}^{n} a_{1k}\Delta FLCF_{t-k} + \sum_{k=0}^{p} a_{2k}\Delta GDP_{t-k} + \\ + \sum_{k=0}^{q} a_{3k}\Delta FDI_{t-k} + \sum_{k=0}^{r} a_{4k}\Delta URB_{t-k} + \lambda_1 FLCF_{t-1} + \lambda_2 GDP_{t-1} \\ + \lambda_3 FDI_{t-1} + \lambda_4 URB_{t-1} + \varepsilon_t\end{aligned} \tag{5}$$

The Fourier terms in the FARDL model capture structural shifts. This represents an advantage compared to ARDL and NARDL models.

3.2 *Kernel-Based Regularized Least Squares (KRLS)*

KRLS is a nonparametric machine learning technique widely recognized for its capability to capture complex, nonlinear relations without assuming any known functional form. Unlike other parametric approaches such as generalized linear models (GLMs), KRLS assumes neither linearity nor additivity effects in empirical analyses [30]. KRLS uses kernel functions to measure the similarity between observations to obtain a smooth and data-driven approximation of the dependence between inputs and outputs.

This feature makes the model robust against overfitting and against the influence of outliers or high-leverage points. KRLS yields estimators with properties such as unbiasedness, consistency, and asymptotic normality [31].

An unknown function $y = f(x)$ is estimated. $f(x)$ is expressed as a weighted combination of kernel functions centered on observed input vectors x_i:

$$f(x) = \sum_{i=1}^{N} c_i k(x, x_i) \tag{6}$$

In Eq. (6), $k(x, x_i)$ represents a kernel function that measures the similarity between the new input x and each observation x_i. The coefficients c_i are determined by regularization. By this approach, the response variable y is not expressed as a direct function of covariates, but rather as a similarity-weighted function of the training data.

Most applications use the Gaussian kernel, also known as the radial basis function (RBF) kernel:

$$k(x_i, x_j) = \exp\left(\frac{-x_i - x_j^{\,2}}{\sigma^2}\right) \tag{7}$$

This kernel function attains its maximum when $x_i = x_j$ and decreases exponentially as the Euclidean distance between x_i and x_j increases. The weights c_i are similar to regression coefficients. Instead of being estimated by ordinary least squares, the weights are computed by Tikhonov regularization [32], which ensures model stability and avoids overfitting.

KRLS constructs a symmetric, positive semidefinite kernel matrix $\boldsymbol{K}$ where each entry is $k(x, x_i)$. The outcome vector $\boldsymbol{y}$ is estimated as follows:

$$y = Kc \tag{8}$$

To ensure a robust solution, the following penalized optimization problem is solved:

$$\min_{f \in H} \sum_{i=1}^{N} \left(y_i - f(x_i)\right)^2 + \lambda f_H^2 \tag{9}$$

The closed-form solution for the coefficient vector $\boldsymbol{c}$ is given by

$$c^* = \arg\min\left[\left(y - Kc\right)^T \left(y - Kc\right) + \lambda c^T Kc\right] \tag{10}$$

In Eq. (10), λ is the regularization parameter that controls the level of penalization to prevent overfitting. Adding λ to the diagonal of $\boldsymbol{K}$ ensures that the matrix is positive definite and invertible [33].

3.3 Fourier-KRLS Model

In the real world, many empirical datasets have smooth structural breaks, cyclical behavior, or long-run nonlinear trends. In such contexts, purely data-driven kernel methods may underperform if they fail to account for temporal patterns explicitly. To solve this issue, several Fourier-based versions of KRLS have been proposed (e.g., [34, 35]). In this study, we integrate Fourier terms into the KRLS model.

The extended Fourier-KRLS regression model is formulated as

$$y_t = f(x_t) + \sum_{k=1}^{K}\left[\alpha_k \sin\left(\frac{2\pi kt}{T}\right) + \beta_k \cos\left(\frac{2\pi kt}{T}\right)\right] + \varepsilon_t \tag{11}$$

In Eq. (11), y_t is the dependent variable, x_t represents covariates, $f(x_t)$ is estimated by means of KRLS, K is the order of the Fourier approximation, and T is the total number of observations.

In practice, K is usually limited to one or two harmonics to avoid overfitting when capturing low-frequency trends [36]. Higher orders (e.g., $K \geq 3$) are justified when datasets contain more frequent fluctuations (e.g., quarterly/monthly observations).

In our application, we use a second-order Fourier expansion ($K = 2$) to approximate smooth policy-driven environmental cycles.

We compute for both KRLS and Fourier-KRLS models the average marginal effects (AMEs) and pointwise marginal effects (PMEs). AMEs are the average of all pointwise marginal effects across the sample. AMEs reflect the overall impact of a predictor. In the Fourier-KRLS model, the AMEs of the sine and cosine terms can be interpreted as the mean influence of unobserved cyclical behavior on the response variable. PMEs represent the estimated effect of a regressor on the dependent variable at each observation, reflecting local heterogeneity.

We compare the AMEs and PMEs for KRLS and Fourier-KRLS models in the empirical part.

4 Empirical Results

4.1 Fourier-ARDL Model

Table 1 contains the variables and their sources. This selection of variables is often used for assessing the influence of environmental and economic factors on fishing sustainability [37, 38].

Table 2 contains the descriptive statistics. The average FLCF is 0.929, showing moderate fishing pressure. The FLCF distribution is normal. The GDP distribution is stable and not normal. It can be interpreted as consistent growth with occasional shocks. The FDI distribution is normal and volatile, indicating irregular investment flows. The URB distribution is normal and has the least variation of 0.048, reflecting constant urban growth during the analyzed period.

We check the data stationarity by means of the Zivot-Andrews unit root test [39] (Table 3), the augmented Dickey-Fuller (ADF) test by [40] (Table 4), and the Fourier augmented Dickey-Fuller (FADF) by Enders and Lee [41].

Table 1 Variable description

Variable	Definition	Measurement unit	Data source
FLCF	Fishing load capacity factor	Index	Global Footprint Network
GDP	Gross domestic product per capita	USD (constant prices, 2015)	World Bank
FDI	Foreign direct investment, net inflows	Percent of GDP (%)	World Bank
URB	Urban population (% of total population)	Percent (%)	World Bank

Source: Authors' collections from various databases

Table 2 Descriptive statistics

	FLCF	GDP	FDI	URB
Mean	0.929	11.143	0.663	4.359
Median	0.971	11.200	0.767	4.360
Maximum	2.152	11.286	1.818	4.435
Minimum	0.168	10.819	−1.346	4.276
Std. Dev.	0.580	0.133	0.807	0.048
Skewness	0.513	−1.182	−0.782	−0.100
Kurtosis	2.318	3.200	3.039	1.874
Jarque-Bera	2.214	8.205	3.571	1.908
Probability	0.330	0.017	0.168	0.385

Source: Author's calculation

Table 3 Zivot-Andrews breakpoint unit root test results

Variables	*T*-statistics	Break year
FLCF	−3.59 (0.342)	2019
GDP	−3.63*** (0.002)	2009
FDI	−5.57*** (0.005)	2014
URB	−4.50 (0.171)	2006

Source: Author's calculation

*, **, *** indicate the significance of variables at 10%, 5%, and 1% levels, respectively

Table 4 ADF unit root test results

	Level	First difference	
Variables	*T*-statistics (Prob)	*T*-statistics (Prob)	Order of integration at 5% L.O.S.
FLCF	−0.40 (0.897)	−7.50*** (0.000)	*I*(1)
GDP	−4.55*** (0.009)	−3.34** (0.020)	*I*(0)
FDI	−4.00*** (0.003)	−8.89*** (0.000)	*I*(0)
URB	−2.52 (0.120)	−4.18***(0.003)	*I*(1)

Source: Author's calculation

*, **, *** indicate the significance of variables at 10%, 5%, and 1% levels, respectively

The Zivot-Andrews unit root test takes into account a structural break within the data. This feature proves to be effective when analyzing time-series data that may undergo abrupt shifts. From Table 3, one can see that GDP and FDI are *I*(0), while FLCF and URB are not stationary.

From Table 4, one can see that FLCF and URB are integrated of order 1 (*I*(1)), while GDP and FDI are stationary in level (*I*(0)).

Table 5 presents the Fourier augmented Dickey-Fuller (FADF) test statistics for both the levels and first differences of the four time series. According to Enders and Lee (2012), for a sample size of $T = 35$ and a trend specification, the critical values are approximately −4.64 (1%), −4.17 (5%), and −3.88 (10%). A variable is considered stationary if its FADF statistic is smaller than the corresponding critical value.

Table 5 Fourier ADF unit root test results

Variables	Level	First difference	Order of integration at 5% L.O.S.
FLCF	−2.508	−4.924***	*I*(1)
GDP	−2.756	−4.289**	*I*(1)
FDI	−3.417	−5.099***	*I*(1)
URB	−3.131	−3.706	*I*(1)

Source: Author's calculation
*, **, *** indicate the significance of variables at 10%, 5%, and 1% levels, respectively

Table 6 VAR lag order selection criteria

Lag	LogL	LR	FPE	AIC	SC	HQ
0	87.65	NA	5.32e−08	−5.39	−5.21	−5.33
1	249.51	271.50	4.41e−12	−14.80	−13.88*	−14.50
2	264.72	21.58	4.91e−12	−14.75	−13.09	−14.21
3	279.95	17.68	6.01e−12	−14.70	−12.30	−13.92
4	313.03	29.88[a]	2.75e−12[a]	−15.80[a]	−12.66	−14.78[a]

Source: Author's calculation
LR sequential modified LR test statistic (each test at 5% level), *FPE* final prediction error, *AIC* Akaike Information Criterion, *SC* Schwarz Information Criterion, *HQ* Hannan-Quinn Information Criterion
a Indicates the lag order selected by the criterion

Table 7 FARDL(4,2,0,3) cointegration bounds test

F-bounds test		Null hypothesis: no cointegration relationship		
Test statistic	Value	Significance (%)	*I*(0)	*I*(1)
F-statistic	6.61	10	2.37	3.2
k	3	5	2.79	3.67
		1	3.65	4.66

Source: Author's calculation

Table 6 shows that four out of the five selection criteria indicate that a lag length of 4 is the most appropriate choice for the FARDL model.

In Table 7, since the computed *F* statistic exceeds the upper bound *I*(1) at all conventional levels, the null hypothesis of no cointegration is rejected. According to AIC, the best model is FARDL(4,2,0,3) (see Fig. 2).

The estimated long-term FARDL coefficients are presented in Table 8. A 1% rise in URB is associated with a 25.94% increase in FLCF in the long term. Urbanization in Norway reduces pressure on marine ecosystems. Urbanization could result in increased consumption of fish products, more intensified coastal development, and environmental pressure from runoff, pollution, or habitat alteration. Unsustainable urban planning could allow urban growth to alter the efforts in sustaining the ecological integrity of marine resources.

A 1% increase in GDP is associated with a 6.92% decrease in FLCF in the long term. Norwegian economic growth can make it possible to increase the fishing

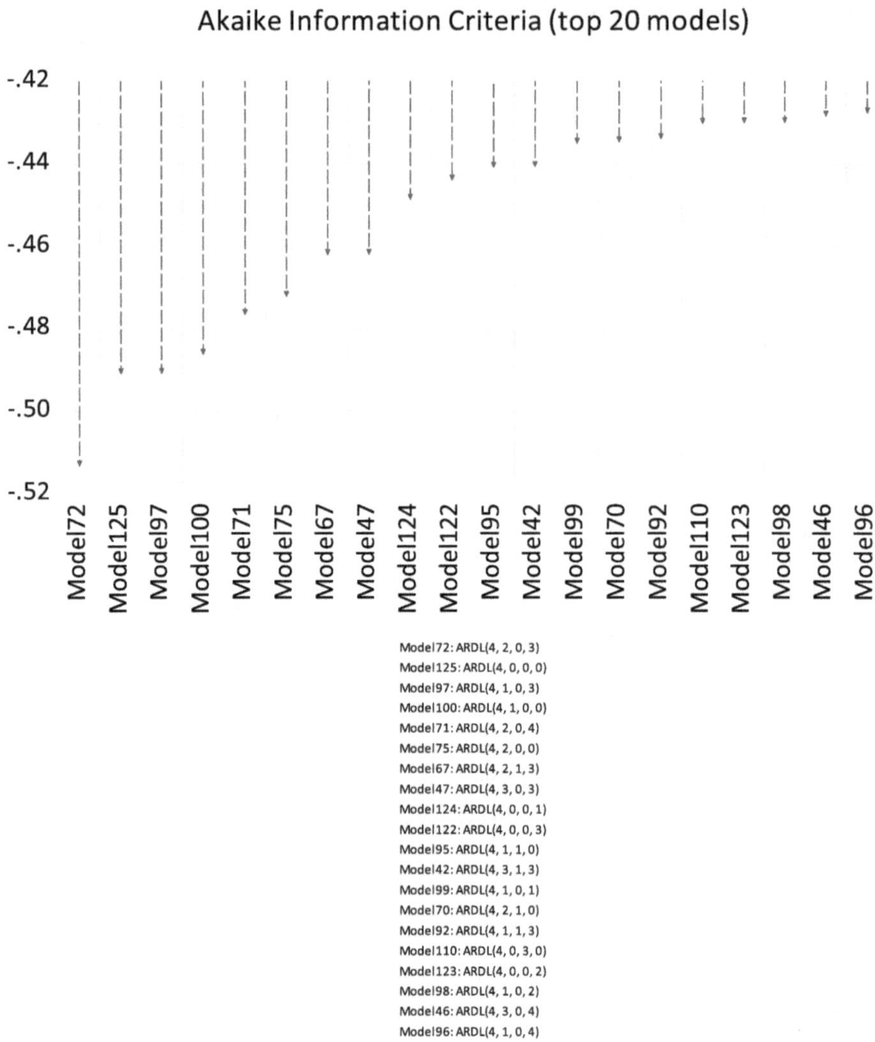

Fig. 2 FARDL model choice based on AIC

pressure due to the enforcement of environmental measures, increased technological advancements within the fishing industry, and increased investment in marine protection.

A 1% rise in FDI is associated with a 0.11% increase in FLCF in the long term. This weak influence implies that FDI might play a small role in Norwegian fisheries-related pressures. This is possible since foreign capital may be used to expand fisheries or related infrastructure without adequate environmental safeguards. These findings align with recent nonlinear analyses of the fishing footprint that emphasize

Table 8 Long-run estimated results of the FARDL model

Variables	Coefficient	*T*-statistics	Probability
GDP	−6.92	−2.94	0.009***
FDI	0.11	2.52	0.022**
URB	25.94	3.38	0.003***
C	−34.95	−3.51	0.002***

Source: Author's calculation
*, **, *** indicate the significance of variables at 10%, 5%, and 1% levels, respectively

Table 9 Short-run estimated results of the FARDL model

Variable	Coefficient	Std. Error	*T*-Statistic	Prob.
D(FLCF(−1))	0.83	0.21	3.89	0.001***
D(FLCF(−2))	1.06	0.20	5.24	0.000***
D(FLCF(−3))	1.07	0.19	5.52	0.0***
D(GDP)	−5.58	2.42	−2.30	0.035**
D(GDP(−1))	3.72	2.25	1.66	0.117
D(URB)	−4.73	46.37	−0.10	0.920
D(URB(−1))	58.25	74.89	0.78	0.448
D(URB(−2))	−107.05	46.60	−2.30	0.036**
COS(2*@PI*0.990000000000001*@ OBSNUM/@OBSSMPL)	−0.43	0.08	−5.64	0**
SIN(2*@PI*0.990000000000001*@OBSNUM/@ OBSSMPL)	−0.09	0.05	−1.71	0.10*
CointEq(−1)*	−1.49	0.23	−6.43	0***
R-squared	0.738081			
Adjusted *R*-squared	0.607122			

Source: Author's calculation
*, **, *** indicate the significance of variables at 10%, 5%, and 1% levels, respectively

the heterogeneous effects of FDI and economic growth across environmental regimes in Norway [42].

The short-run dynamics of FLCF in Table 9 proves that FLCF is significantly impacted by its own past values and the other predictors. The first three lags of the differenced dependent variable FLCF are all positive and statistically significant, indicating strong short-term inertia in FF. Past increases in fishing pressure tend to persist in the short term.

A 1% increase in GDP is associated with a 5.58% reduction in FLCF. FLCF can be diminished as a result of stringent environmental laws. The lagged GDP effect is positive but not statistically significant.

D(URB) and D(URB(−1)) are statistically insignificant. D(URB(−2)) is negative and statistically significant. This effect reflects the delay in which coastal development policy or urban infrastructure could have an impact on the marine ecosystem.

The negative coefficient of the cosine Fourier term indicates smooth cyclical patterns due to seasonal or policy-induced patterns. The sine term is on the verge of marginal significance, indicating a potential weaker cyclical pattern in addition to the more dominant cosine effect.

The ECT is −1.49, indicating that about 149% of any short-term deviation from the long-run equilibrium is corrected the following year. This points to a very fast speed of adjustment.

The *R*-squared and the adjusted *R*-squared indicate that 60% of short-run variation in FLCF is explained by the regressors.

The test results and null hypotheses are shown in Table 10. The model's normality, lack of autocorrelation, and robustness are confirmed since the *p*-values exceed 5%. The CUSUM plot in Fig. 3 stays within the 5% significance bounds and validates the model's stability.

Table 10 Diagnostic results for the FARDL model

Diagnostic test	H_0	Decision statistics [*p*-value]
χ^2 SERIAL	There is no serial correlation in the residuals	Accept H_0 2.43[0.126]
χ^2 ARCH	There is no autoregressive conditional heteroscedasticity	Accept H_0 0.07[0.781]
χ^2 Jarque Bera	Normal distribution	Accept H_0 2.32[0.313]
χ^2 Ramsey	Absence of model misspecification	Accept H_0 1.88[0.187]

Source: Author's calculation

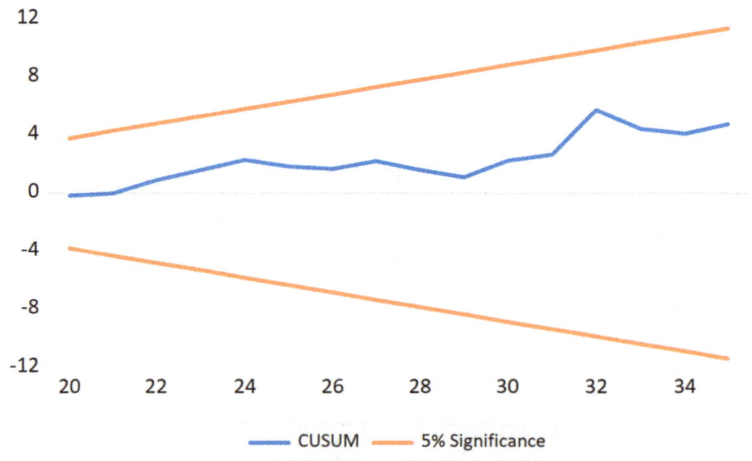

Fig. 3 ARDL plot of CUSUM test. (Source: Author's calculation)

4.2 KRLS and Fourier-KRLS Comparison

In this section, we compute and compare the AMEs and PMEs for both standard KRLS and Fourier-KRLS models. Table 11 presents the AMEs estimated using the standard KRLS model. One can notice that URB has the strongest and statistically significant marginal effect on FLCF. This reflects the role of urban infrastructure in complementing environmental monitoring, regulation, or management of resources. GDP has a positive and statistically weaker impact (AME = 0.84), with a wide confidence interval including zero. This means that economic growth is not always responsible for FLCF variation. FDI has a very low average marginal effect (AME = 0.05), which is not statistically significant.

Table 12 presents the AMEs from the Fourier-KRLS model, which includes the Fourier terms. They account for multiple seasonal patterns. In this model, URB has a positive and statistically significant effect on FLCF, of a lower magnitude. Both FDI and GDP record negligible marginal effects, consistent with the conclusion that macroeconomic variables on their own do not directly contribute to FLCF fluctuations. The Fourier terms (cos_1 = 20.18, cos_2 = 5.06) are high and highly significant. They confirm the presence of cyclical or seasonal patterns, caused by fish migration or seasonal fishing regulations.

In the Fourier-KRLS model, URB has a positive, statistically significant effect on FLCF, with a lower impact than in the KRLS model. GDP and FDI both have small, nonsignificant marginal effects. This leads to the conclusion that macroeconomic variables taken alone do not directly explain FLCF variations. The large values of the Fourier terms (e.g., cos_1 = 20.18, cos_2 = 5.06) are highly significant, indicating the presence of strong cyclical or seasonal patterns in FLCF.

Table 11 Average marginal effects (AME)—standard KRLS model

Variable	AME	SE	CI lower	CI upper
GDP	0.8424	0.6467	−0.4251	2.1099
FDI	0.0542	0.6467	−1.2134	1.3217
URB	9.0459	0.6467	7.7784	10.3134

Source: Author's calculation

Table 12 Average marginal effects (AMEs)—Fourier-KRLS model

Variable	AME	SE	CI lower	CI upper
GDP	0.2535	0.5126	−0.7513	1.2581
FDI	0.0932	0.5126	−0.9115	1.0978
URB	1.5919	0.5126	0.5873	2.5967
sin_1	4.6159	0.5126	3.6113	5.6206
cos_1	20.1814	0.5126	19.1767	21.1861
sin_2	2.3279	0.5126	1.3232	3.3326
cos_2	5.0573	0.5126	4.0529	6.0622

Source: Author's calculation

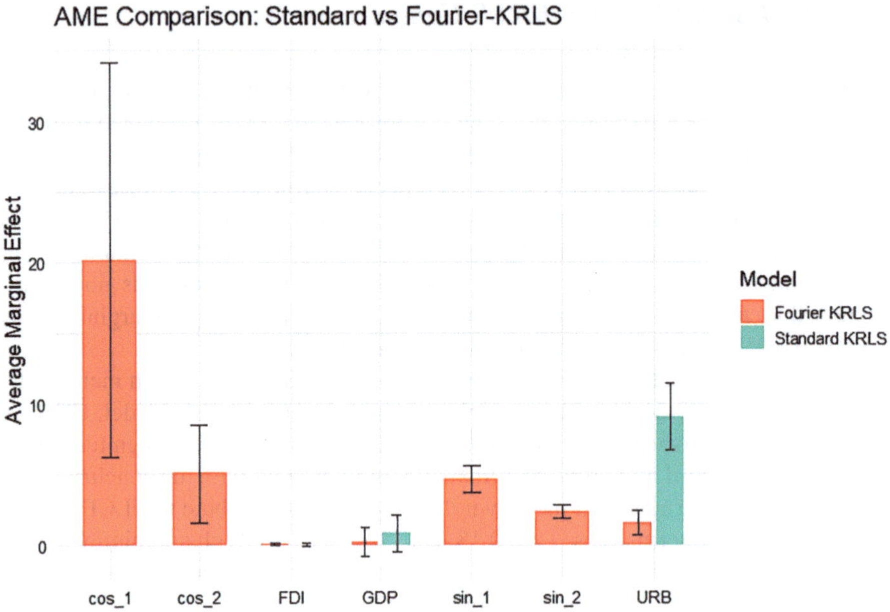

Fig. 4 Comparison of AMEs from KRLS and Fourier-KRLS models on Norway's fishing footprint. (Source: Author's calculation)

Table 13 Distribution of PMEs in the standard KRLS model

Variable	25th Percentile (P25)	Median (P50)	75th Percentile (P75)
GDP	0.198	0.675	1.960
FDI	−0.126	0.085	0.164
URB	7.260	11.400	12.500

Source: Author's calculation

Figure 4 shows a visual comparison of AMEs for the Fourier-KRLS and KRLS models. In the Fourier-KRLS model, the cosine and sine terms have the strongest effects, especially cos_1.

This reinforces the idea that cyclical or seasonal patterns explain FLCF.

Table 13 reports the distribution of PMEs for FLCF predictors in the case of the KRLS model. PMEs reflect the variation of the marginal effect of each variable across individual observations in the sample. URB has the strongest and most consistent positive effect on FLCF, with a median marginal effect of 11.4 and a narrow interquartile range. URB is associated with a higher FLCF. This direct URB-FLCF relationship can be brought on by a higher demand for marine resources. The positive impact of GDP on FLCF is moderate. The median PME is 0.675, but the range of values suggests a varying GDP-FLCF relationship.

Table 14 Distribution of PMEs in the Fourier-KRLS model

Variable	P25	P50	P75
GDP	−0.462	−0.226	0.900
FDI	0.0501	0.119	0.175
URB	−0.119	1.78	2.43
sin_1	1.55	5.13	6.56
cos_1	−20.9	−5.59	66.7
sin_2	0.777	2.57	3.31
cos_2	−5.27	−1.41	16.7

Source: Author's calculation

FDI effect on FLCF is weak. Its 25th percentile is negative, and its median tends to be zero. Depending on the conditions or timing, FDI can slightly increase or decrease FLCF.

Table 14 presents the PMEs at the 25th, 50th, and 75th percentiles of the Fourier-KRLS model. At the lower and median percentiles, the effects of GDP on FLCF are negative. At the upper quantile, the effect becomes positive. This suggests a nonlinear or threshold effect. It means that economic development is not uniformly beneficial to the fishing industry, and it can exert downward pressure on FLCF in the early stages, supporting it at higher development levels.

FDI has a positively consistent pattern in all percentiles. This indicates that FDI complements FLCF by means of improved technology, infrastructure, or management efficiency. Even if the marginal effect of FDI is small, its upward trend points to an intensification of its effects when investments are implemented into the domestic fishing sector.

URB has a varying influence. It is negative in the 25th percentile, positive at the median, and increases in the 75th percentile. When URB is low, the fishing industry is restricted by underdeveloped infrastructure. When URB rises, its effect becomes positive due to better logistics or urban coastal economies that favor fishing operations.

The Fourier terms show high and oscillating effects. Sin_1 has a steady positive effect (from 1.55 to 6.56), proving that seasonal cycles increasingly affect FLCF. Cos_1 exerts a wide range of effects, from a substantial negative value −20.9 at P25 and −5.59 at P50 to a high positive effect of 66.7 at P75. This variation suggests nonlinear cyclical dynamics. The second-order Fourier terms, sin_2 and cos_2, encompass shorter cyclical factors. The marginal effects of sin_2 are uniformly positive and relatively steady, ranging from 0.777 to 3.31. These short cycles favor FLCF. In lower percentiles, sin_2 has lower negative values, rising to higher positive values at P75.

Figure 5 visualizes the dynamic PMEs of FDI, GDP, and URB on FLCF for KRLS (blue) and Fourier-KRLS (red). The marginal effects of FDI are positive and very stable in the Fourier cases, while in the standard case, the effects are volatile with changing signs. The cyclical component of the Fourier model smoothens the FDI effect and creates a positive, steadier relation between FDI and FLCF.

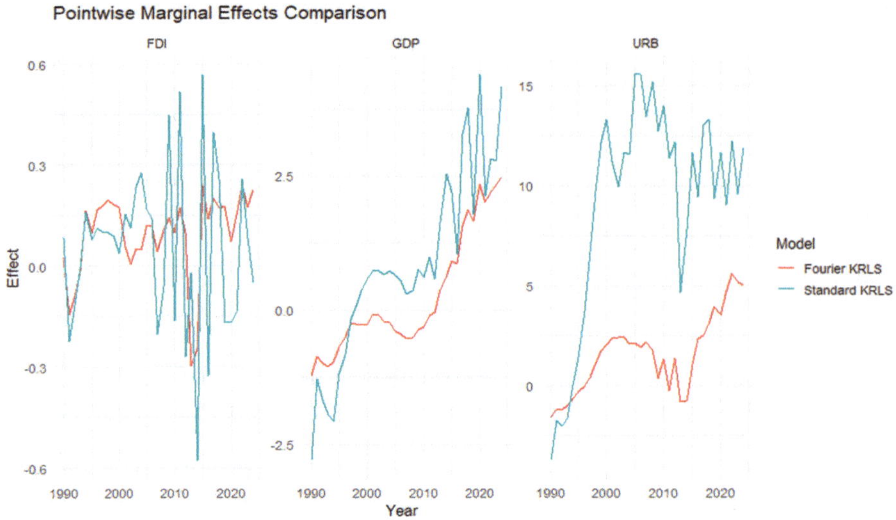

Fig. 5 Temporal evolution of pointwise marginal effects on FLCF: a comparison between standard and Fourier-KRLS models. (Source: Author's calculation)

In the standard case, one can identify a highly volatile and asymmetric impact of GDP on FLCF, with a sharply rising trend after 2010. The Fourier model shows a smoother surge since the beginning of the 2000s. In later years, economic growth supports FLCF, but the Fourier-KRLS model provides more trustworthy results regarding its magnitude.

In the standard model, URB marginal effects are very large, reaching a peak of around 2005–2010. The Fourier-KRLS estimates are smooth and imply a smooth shift from negative to positive values. When we include cyclical components in the model, URB's effect on FLCF changes more systematically and less randomly. From this analysis, one deduces that Fourier components lead to smoother and less volatile trajectories over time.

Figure 6 plots the actual FLCF values and the estimated values from the two models. The black solid line represents actual FLCF values. The blue dashed line and green dotted line symbolize the fitted values from the standard and Fourier models. The Fourier-KRLS line follows the actual path, especially at turning points and when the growth becomes steeper after 2010. It implies that the Fourier-KRLS model has a higher ability to follow the actual FLCF trend. The standard KRLS line diverges more when it comes to abrupt changes (early 2000s and 2020), showing potential underfitting.

Table 15 shows that the Fourier-KRLS model slightly outperforms the standard KRLS model in terms of predictive accuracy and goodness of fit. The Fourier-KRLS model has a lower mean absolute percentage error (MAPE) and root mean square error (RMSE) compared to the standard model, and a higher R^2.

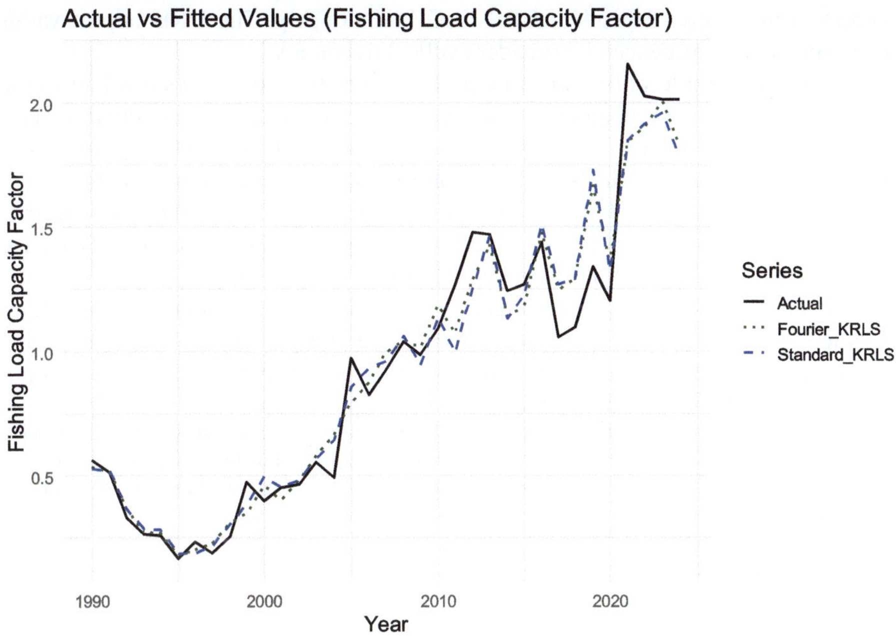

Fig. 6 Actual vs. fitted fishing load capacity factor: comparison of standard and Fourier-KRLS models (1990–2024). (Source: Author's calculation)

Table 15 Accuracy metrics comparison between standard KRLS and Fourier-KRLS models

Model	MAPE	RMSE	R^2
Standard KRLS	0.108971	0.134042	0.945043
Fourier KRLS	0.107598	0.125764	0.951621

Source: Author's calculation

5 Conclusion and Policy Recommendations

This study investigates the drivers of Norway's FLCF during 1990–2024 using a Fourier-based approach: Fourier-KRLS and FARDL. The Fourier components proved the nonlinear and cyclic relationship between FLCF and its drivers.

Hypothesis H1 is confirmed, since GDP per capita exhibits a negative long-run effect on FLCF. This suggests that economic growth raises fishing pressure and reduces the sustainability of marine ecosystems.

Hypothesis H2 is partially supported, since FDI has a weak positive influence on FLCF. It follows that capital inflows play a limited beneficial role in sustaining marine ecosystems.

Hypothesis H3 is partially supported, since URB shows a positive long-run effect on FLCF in the FARDL model. However, the Fourier-KRLS results reveal a nonlinear and phase-dependent URB-FLCF relationship. The URB impact varies over time and is influenced by broader cyclical dynamics.

Several actionable policies can be derived from this study. Smart urbanization strategies should be implemented to support environmental sustainability of marine resources. Second, FDI should be channeled to sustainable aquaculture, low-impact fishing methods, and marine biodiversity preservation. Third, we obtained that economic growth reduces FLCF. The economic progress should be balanced with strict fish stock management to avoid exceeding biocapacity limits. Fourth, responsive fishing activity should be monitored by real-time monitoring systems.

The study also has some limitations. Seasonal dynamics peculiar to fishery management can be better reflected by monthly or quarterly data. Illegal or unreported fishing practices may not be fully reflected in the FLCF indicator. Fourier components might not capture environmental shocks or policy shifts.

Future research could be directed toward using higher-frequency datasets. A spatial econometric approach could allow comparisons between countries or regions with similar fishing patterns. Agent-based systems are another way by which we could potentially model the long-term effects of different policy choices.

Statements and Declarations

Funding The authors declare that no funds, grants, or other support were received during the preparation of this manuscript.

Competing Interests The authors have no relevant financial or non-financial interests to disclose.

Ethics Approval and Consent to Participate The authors declare that they have no known competing financial interests or personal relationships that seem to affect the work reported in this article. We declare that we have no human participants, human data, or human tissues.

Consent for Publication All of the authors have read and approved the paper for publication.

Author Contributions All authors contributed to the study conception and design. Material preparation, data collection, and analysis were performed by all authors. The first draft of the manuscript was written by authors, and all authors commented on previous versions of the manuscript. All authors read and approved the final manuscript.

Data Availability The data that support the findings of this study are available from the corresponding author upon reasonable request.

AI Use Disclosure Statement The authors did not use any generative AI or AI-assisted technologies in the creation of this manuscript. The authors used generative AI technologies only for proofreading. This includes minor corrections to grammar, spelling, and syntax. No AI was used for ideation, data analysis, or writing original content.

References

1. Adalı Z, Toygar A, Yıldırım U (2023) Assessing the stochastic behavior of fishing grounds footprint of top ten fishing countries. Reg Stud Mar Sci 63. https://doi.org/10.1016/j.rsma.2023.103015
2. Wackernagel M, Rees WE (1996) Our ecological footprint: reducing human impact on the earth. New Society Publishers. https://books.google.ro/books/about/Our_Ecological_Footprint.html?id=N__ujKDfXq8C&redir_esc=y
3. Kitzes J, Peller A, Goldfinger S, Wackernagel M (2007) Current methods for calculating national ecological footprint accounts. Sci Environ Sustain Soc 4(1):1–9. https://elearning.humnet.unipi.it/pluginfile.php/101791/mod_resource/content/0/FootprintMethod2006.pdf
4. Pata UK, Caglar AE (2021) Investigating the EKC hypothesis with renewable energy consumption, human capital, globalization and trade openness for China: evidence from augmented ARDL approach with a structural break. Energy 216. https://doi.org/10.1016/j.energy.2020.119220
5. Georgescu I, Kinnunen J (2024) Effects of FDI, GDP and energy use on ecological footprint in Finland: an ARDL approach. World Dev Sustain 4. https://doi.org/10.1016/j.wds.2024.100157
6. Galli R (1998) The relationship between energy intensity and income levels: forecasting long term energy demand in Asian emerging countries. Energy J 19(4):85–105. https://doi.org/10.5547/ISSN0195-6574-EJ-Vol19-No4-4
7. Siche R, Pereira L, Agostinho F, Ortega E (2010) Convergence of ecological footprint and emergy analysis as a sustainability indicator of countries: Peru as case study. Commun Nonlinear Sci Numer Simul 15(10):3182–3192. https://doi.org/10.1016/j.cnsns.2009.10.027
8. Pata UK, Karlilar Pata S (2024) Determining the effectiveness of the forest load capacity factor in assisting decarbonization in India. For Policy Econ 166. https://doi.org/10.1016/j.forpol.2024.103281
9. Kinnunen J, Georgescu I, Nica I (2025) An evaluation of the load capacity curve hypothesis for Finland: an ARDL approach with frequency domain analysis. Manag Environ Qual 36(5):1233–1260. https://doi.org/10.1108/MEQ-06-2024-0265
10. Yilanci V, Cutcu I, Cayir B (2022) Is the environmental Kuznets curve related to the fishing footprint? Evidence from China. Fish Res 254. https://doi.org/10.1016/j.fishres.2022.106392
11. Sala E et al (2021) Protecting the global ocean for biodiversity, food and climate. Nature 592(7854):397–402. https://doi.org/10.1038/s41586-021-03371-z
12. Sarkodie SA, Owusu PA (2023) Assessment of global fish footprint reveals growing challenges for sustainable production and consumption. Mar Pollut Bull 194. https://doi.org/10.1016/j.marpolbul.2023.115369
13. Jacob C et al (2023) A two way process—social capacity as a driver and outcome of equitable marine spatial planning. Mar Policy 149. https://doi.org/10.1016/j.marpol.2023.105507
14. Singh S, Sharma GD, Radulescu M, Usman M, Balsalobre-Lorente D (2025) Combined asymmetric influences of renewable energy consumption and categorical economic policy uncertainty on economic growth in Japan: new insights from QQR and KRLS approaches. Environ Dev Sustain. https://doi.org/10.1007/s10668-025-06156-w
15. Mondal R, Bhat A (2021) Comparison of regression-based and machine learning techniques to explain alpha diversity of fish communities in streams of central and eastern India. Ecol Indic 129. https://doi.org/10.1016/j.ecolind.2021.107922
16. Yilanci V, Cutcu I, Cayir B, Saglam MS (2023) Pollution haven or pollution halo in the fishing footprint: evidence from Indonesia. Mar Pollut Bull 188. https://doi.org/10.1016/j.marpolbul.2023.114626
17. Wirajing MAK, Nanfosso RT, Kountchou AM (2025) Examining the environmental tax and the fishing grounds footprint nexus: case of the African fishing industry. Nat Res Forum 49(3):2130–2154. https://doi.org/10.1111/1477-8947.12411

18. Mance D, Miljenović D, Velić I (2025) Do economies recover their fisheries? Evidence of an environmental Kuznets curve for fish stock status. Sustainability (Switzerland) 17(14). https://doi.org/10.3390/su17146646
19. Mustafa K, Muhammet Y, Şerif C, Nurullah A (2025) Restoring balance: the intersection of management and fishing sustainability in Indonesia. Fish Manag Ecol. https://doi.org/10.1111/fme.70007
20. Georgescu I, Nica I, Kinnunen J (2024) Towards sustainability: understanding Norway's ecological footprint through the framework of the environmental Kuznets curve. Energies (Basel) 17(23). https://doi.org/10.3390/en17236074
21. Canbay Ş, Şengül S, Karahan-Dursun P (2025) From capture to culture: the asymmetric impact of education on Türkiye's fisheries sector. Aquacult Int 33(6). https://doi.org/10.1007/s10499-025-02153-y
22. Obekpa HO, Alola AA, Adejo AM, Echebiri C (2025) Examining the drivers of ecological footprint components: is pursuing food security environmentally costly for Nigeria? Ecol Indic 170. https://doi.org/10.1016/j.ecolind.2024.113009
23. Solarin SA, Lafuente C, Gil-Alana LA, González-Blanch MJ (2025) Fishery products consumption: evidence of persistence and trends in 25 OECD countries. Aquac Fish. https://doi.org/10.1016/j.aaf.2025.07.003
24. Georgescu I, Kinnunen J (2025) Nonlinear effects of GDP regimes, renewable energy, and urbanization on Finland's ecological footprint: an MT-NARDL approach. World Dev Sustain 7. https://doi.org/10.1016/j.wds.2025.100235
25. Lütkepohl H (2005) New introduction to multiple time series analysis. Springer Berlin Heidelberg, Berlin. https://doi.org/10.1007/978-3-540-27752-1
26. Pesaran MH, Shin Y, Smith RJ (2001) Bounds testing approaches to the analysis of level relationships. J Appl Econ 16(3):289–326. https://doi.org/10.1002/jae.616
27. Yilanci V, Bozoklu S, Gorus MS (2020) Are BRICS countries pollution havens? Evidence from a bootstrap ARDL bounds testing approach with a Fourier function. Sustain Cities Soc 55. https://doi.org/10.1016/j.scs.2020.102035
28. Ludlow J, Enders W (2000) Estimating non-linear ARMA models using Fourier coefficients. Int J Forecast 16(3):333–347. https://doi.org/10.1016/S0169-2070(00)00048-0
29. Enders W, Lee J (2012) A unit root test using a Fourier series to approximate smooth breaks. Oxf Bull Econ Stat 74(4):574–599. https://doi.org/10.1111/j.1468-0084.2011.00662.x
30. Hainmueller J, Hazlett C (2014) Kernel regularized least squares: reducing misspecification bias with a flexible and interpretable machine learning approach. Polit Anal 22(2):143–168. https://doi.org/10.1093/pan/mpt019
31. Ferwerda J, Hainmueller J, Hazlett CJ (2017) Kernel-based regularized least squares in R (KRLS) and Stata (KRLS). J Stat Softw 79. https://doi.org/10.18637/jss.v079.i03
32. Tikhonov A, Arsenin V (1977) Solutions of ill-posed problems. Wiley, Washington, DC
33. Schölkopf B (2002) Learning with kernels. In: Proceedings of 2002 international conference on machine learning and cybernetics. https://doi.org/10.7551/mitpress/4175.001.0001
34. Bouboulis P, Pougkakiotis S, Theodoridis S (2016) Efficient KLMS and KRLS algorithms: a random Fourier feature perspective. In: IEEE workshop on statistical signal processing proceedings. IEEE Computer Society. https://doi.org/10.1109/SSP.2016.7551811
35. Xu X, Ren W (2022) Random Fourier feature kernel recursive maximum mixture correntropy algorithm for online time series prediction. ISA Trans 126:370–376. https://doi.org/10.1016/j.isatra.2021.08.014
36. Adebayo TS, Özkan O, Eweade BS (2024) Do energy efficiency R&D investments and information and communication technologies promote environmental sustainability in Sweden? A quantile-on-quantile KRLS investigation. J Clean Prod 440:140832. https://doi.org/10.1016/j.jclepro.2024.140832
37. Ayad H (2023) Investigating the fishing grounds load capacity curve in G7 nations: evaluating the influence of human capital and renewable energy use. Mar Pollut Bull 194:115413. https://doi.org/10.1016/j.marpolbul.2023.115413

38. Emre Caglar A, Gökçe N, Balsalobre-Lorente D (2024) Towards the vision of going green: the role of different energy research and development investments, urbanization and income in load capacity factor. Sustain Energy Technol Assess 68:103888. https://doi.org/10.1016/j.seta.2024.103888
39. Zivot E, Andrews DWK (1992) Further evidence on the great crash, the oil-Price shock, and the unit-root hypothesis. J Bus Econ Stat 10(3):251. https://doi.org/10.2307/1391541
40. Dickey DA, Fuller WA (1979) Distribution of the estimators for autoregressive time series with a unit root. J Am Stat Assoc 74(366):427. https://doi.org/10.2307/2286348
41. Enders W, Lee J (2012) The flexible Fourier form and Dickey–Fuller type unit root tests. Econ Lett 117(1):196–199. https://doi.org/10.1016/j.econlet.2012.04.081
42. Georgescu I, Kinnunen J (2025) Nonlinear impacts of FDI and GDP on the fishing footprint in Norway: insights from quantile-on-quantile and KRLS models. Quality & Quantity. https://doi.org/10.1007/s11135-025-02524-w

A Computational Intelligence Framework for Urban Water Security: Integrating Digital Twins and Hybrid AI for Arid Climates

Uneb Gazder, Md. Kamrul Islam, Nidal Abdul Rahim Dahman, and Monjur Mourshed

Abstract A significant issue facing the Gulf Cooperation Council (GCC) nations is water scarcity. The causes are rapid urbanization, dwindling freshwater supplies, and arid conditions. The water management system for Bahrain suggested by this study incorporates desalination, stormwater harvesting, aqua harvesting, and permeable pavements. Water scarcity is a major challenge for the Gulf Cooperation Council (GCC) countries. It is caused by arid climates, limited freshwater resources, and rapid urban growth. This study proposes a water management framework for Bahrain that combines permeable pavements, stormwater harvesting, aqua harvesting, and desalination. This approach would integrate digital twin technology for real-time water system modeling with hybrid machine learning models. The study is based on a review of published research, selected case studies, and an examination of existing technologies in the region. The development of the framework initially emphasized both qualitative and quantitative aspects of water provision. Quantum computing was suggested as a tool to enhance the supply of safe water. The qualitative part of the framework focuses on optimizing the location of sensors in pipelines for monitoring water quality. In the quantitative part, the framework employs graph neural networks (GNNs) for network optimization, transformer models for demand forecasting, and reinforcement learning for adaptive control systems. For optimization challenges, the framework utilizes advanced algorithms, including metaheuristic approaches for location optimization, and stochastic opti-

U. Gazder · N. A. R. Dahman
Department of Civil Engineering, College of Engineering, University of Bahrain, Sakhir, Bahrain
e-mail: ugazder@uob.edu.bh; ndahman@uob.edu.bh

M. K. Islam (✉)
Department of Civil and Environmental Engineering, College of Engineering, King Faisal University, Al-Ahsa, Saudi Arabia
e-mail: maislam@kfu.edu.sa

M. Mourshed
School of Engineering, Cardiff University, Cardiff, UK
e-mail: MourshedM@cardiff.ac.uk

M. Mirrashid et al. (eds.), *Optimization and Adaptive Control Algorithms*, Emerging Trends in Mechatronics, https://doi.org/10.1007/978-981-95-8387-4_2

mization methods for handling uncertainty. The framework is designed to address several challenges linked to water security and aims to address key challenges in water security. An artificial intelligence-based digital twin model can support this work by handling microlevel models that involve many variables. An AI-based digital twin model supports this framework by handling complex, multivariable micro-level models. The development and implementation of such a framework are considered vital for Bahrain and other arid regions, including neighboring countries like Saudi Arabia.

Keywords Digital twin · Water security · Optimization · Urbanization · Permeable pavements

1 Introduction

Climate change is a well-known phenomenon, which has persisted over a long period of time. One of the immediate implications of this issue is the impact on global water systems, leading to the concept of water security. In broad terms, water security can be considered as a tolerable water-related risk to society [1]. More specifically, this issue can be divided into three main dimensions, which are decline of surface water, depletion of groundwater resources, and degradation of water quality [2]. The water needs can be divided into two categories: green water, which relates to particular food production, and blue water, which caters for human health, energy supply, industrial production, irrigation, and other similar applications [1].

The impact of water scarcity has resulted in varying effects on different geographical regions. These effects can sometimes diversify even within a country depending upon its geographical extents, as reported in a study for China [2]. Its consequences have been diverse including human migration patterns, population growth, agricultural activities, economic development, river flows and geographies, and global environment [3]. Water scarcity is also the major contributor to the issue of food security due to agricultural demand for water. This demand keeps on increasing with the rise in global population [4]. This issue has been found to equally affect the rain-fed as well as irrigated systems of agriculture [5]. Economic development is another factor which plays a role in the need for expanding supply of freshwater resources [6]. Another contributing factor to the unequal distribution of water scarcity issue is the trend of brazen water grab, which is boosted by social and political inequalities [7]. This aspect is covered under the approaches of political ecology and water insecurity, with the latter having impacts on individual behavior. Other approaches to tackle water security and scarcity issues include economic anthropology, human awareness, and responsible behavior. Economic anthropology deals with institutional and financial measures to mitigate water security, while human awareness deals with this matter through individual and community involvement [8].

Water scarcity is one of the most pressing challenges faced by the Middle East countries and those which are part of the Gulf Cooperation Council (GCC), driven by arid climates, limited freshwater resources, and rapid urbanization [9]. The issue of water scarcity in this region is not newly found; in fact, it was initiated in the colonial era with the political grievances of that time period [10]. In addition to the Gulf region, African nations are also under the threat of water security as they face the same issues. Due to perhaps even more critical characteristics and geographical proximity, these countries are termed under the bloc name of the Middle East and North Africa (MENA). The region faces a complex environmental and geopolitical landscape, which aggravates environmental issues, the most important of which is water security. These countries, along with the South Asian subcontinent, are known to have water management issues with extended river-basin overexploitation [6]. Due to these multisectoral impacts, there has been a strong correlation reported between water scarcity and poverty, especially for the sub-Saharan African nations [1].

The need to adopt an integrated and technology-driven approach to solve water security issues is strongly advocated in the literature [5]. The combination of several approaches for water security has been considered in the past in some regions, including policy and technical measures [10]. However, recent studies still point to the development of multisectoral initiatives, especially for the MENA region. To this end, integrated water management, resilient infrastructure, and regional and international collaboration have been identified as possible areas of advancements [11]. Considering this aspect, and to address this critical issue, an innovative water management framework is proposed that integrates the sustainable options of permeable pavements, stormwater and aqua harvesting, and desalination with the power of quantum computing to ensure meeting the water demands of Bahrain. The modern concept of "digital twin" application is recommended for accurate and efficient provision of safe water. Thus, this study suggests a multisectoral, multifaceted, all-encompassing strategy for addressing this problem by utilizing computing technology breakthroughs. Using robust infrastructure and integrated management, the framework tackles the problem of water security. The unmatched processing power that quantum computing provides to analyze intricate datasets improves this methodology.

A major element of the proposed framework focuses on optimizing the location of sensors in the pipelines for monitoring water quality. From the quantitative part, the framework suggests using transformer models for forecasting demand at the microlevel taking into account the previous trends as well as the demographic characteristics of the area. For the other part of the quantitative division, graph neural networks (GNNs) are to be used for network optimization including locations of desalination plants, permeable pavements, aqua harvesting stations, and water reservoirs and their integration with each other and the current supply system.

The research was initiated with a detailed literature review, study of use cases of artificial intelligence (AI)-based computing, and then exploration of the existing technologies which have been currently developed for this arid region. The development of the framework started with a focus on qualitative and quantitative aspects of water provision. The application of the digital twin model was recommended for accurate and efficient provision of safe water.

2 Background

2.1 Water System Monitoring

Monitoring and ensuring the confirmation of standards with regard to quality of water reaching the citizens are both a national and global requirement set by the World Health Organization (WHO) [12]. The monitoring of drinking water and sanitation has been recognized as a global issue. Consequently, it is made part of the Millennium Development Goals (MGDs) [13]. National strategies play a vital role in ensuring sustainable supply of safe and healthy drinking water. There is a need to deploy development-oriented strategies in this regard. Such strategies are expected to have far-reaching long-term impacts on the socio-economic conditions of the country [14].

The issue becomes more critical in arid and semiarid regions due to the high cost of water supply and scarcity of available sources. The water supply system is one of the most critical and expensive parts of the national infrastructure. The performance indicators for such systems comprise a wide variety of factors including physical assets, staffing, operations, customer satisfaction, and economics. These indicators may differ according to the size of the system. Among these indicators, those related to water quality and environmental aspects have traditionally been given lower priority compared to indicators related to physical assets, operation, and cost. The progress of research has gone to the extent of using mobile device-based technologies with extended communication systems to detect pipe damage and defects [15]. However, the situation is changing in the recent decade or so with more attention given to safe and healthy water supply and environmental impacts [16]. The environmental impacts of the water supply are linked to the water supply system as well as the energy consumption on the consumer side. Hence, various algorithms have been implemented for smart energy utilization for the consumer using sensors and controllers [17].

Technological advances have been playing their role in increasing the efficiency of infrastructure design, maintenance, management, and rehabilitation. A smart monitoring and management system would allow the authorities to solve and mitigate the problems proactively and promptly. An automated platform for water system monitoring would comprise monitoring architectures including real-time supervision and control components, logic controllers equipped with common function libraries, communication systems, and interfaces between the controllers and the employed monitoring, electrical, and mechanical systems. Such integrated systems are commonly referred to as Supervisory Control and Data Acquisition (SCADA) [18]. The traditional methods of water management come at a high cost and incur a considerable amount of energy demand [19].

In this regard, the concept of the Internet of Things (IoT) has emerged as an integrated solution to smart data-driven decision-making. This system enables real-time and continuous monitoring of the water distribution system. It has been used for monitoring specific elements of water distribution. These elements include power requirements and water pressure [20]. The effective and successful implementation of such an advanced system requires carefully installed sensors, a controller to handle the analysis and decision-making, and cloud space for the open availability of data for processing. Some studies have suggested the use of a Raspberry Pi system in IoT-based monitoring systems [21]. However, the application of IoT has been more linked to the management of the infrastructure rather than water quality. There has been more than one example of this sort of application of the automated systems in the literature, such as by Kim et al. [22], which focuses on water quantity rather than quality. Related to the application of IoT in water monitoring systems, it has been found that the accuracy of the models relies on suitable and sufficient datasets. In this context, previous studies have employed a number of techniques including support vector networks, K-nearest neighbor, and Random Forest [23]. There is still the need to advance modeling capabilities in this field with the inclusion of state-of-the-art techniques [24]. Quantum computing could be a promising candidate to move forward in this area. But this does not entail that the sensor technology for water quality monitoring is not available. Stoianov et al. [25] have shown the application of wireless sensors for water quality monitoring. Their proposed system was used successfully as a prototype on water supply and sewer systems to monitor water quality and other hydraulic parameters such as pressure and flow. Nondestructive sensing of pipe conditions has also been successfully applied and modeled to provide advance detection of pipe damage [26]. The sensor data is sampled and analyzed on a periodic basis (weekly/biweekly/monthly) for detecting anomalies related to water quality. According to Liu et al. [27], this type of detection system consists of gathering information about the different physical and chemical characteristics of water and evaluating it to provide indicators of water quality. These characteristics, which are generally acknowledged as significant markers of water quality, include temperature, turbidity, pH, and total dissolved solids [28]. Conductivity, dissolved oxygen (DO) level, and pH are a few other pertinent factors [21], illustrating the wide range of factors that go into assessing water quality. The placement of sensors, and the network protocol, is another major issue which has been identified as a point of concern to effectively manage the water management system. This is true for the case of water quality as well as quantity. The effective implementation of IoT cannot be ensured without addressing such issues properly [29].

However, there is still a lot of scope for research in this area to understand the chemical and physical processes that affect water quality during distribution and transmission processes.

2.2 Water Demand Prediction

The current system of water supply management relies on the reactive approach of demand management and monitoring of water quality from the source or the supply end [30]. Accurate prediction of water demand and proactive planning for it require information related to consumption, production, and resources. Most of the authorities rely on the water-meter readings to monitor the consumption. Such a system, even with the required accuracy, does not provide the spatial granularity to provide an overall view of the consumption cycle and needs [22].

Furthermore, there is the impact of temporal variations in water demand. Such variations could be misjudged due to the sampling frequencies. A lack of comprehension of such variations and details could be damaging to the network performance [31].

Moreover, there is virtually no role of the latest technologies such as rainwater, aqua harvesting, and permeable pavements in meeting the water demand [32]. The water supply demand is generally managed using desalination in most of the GCC countries [33] and other treatment methods of available water, which involve large investments and lack variety in sources, making it susceptible to water security issues.

The problem with adopting a comprehensive approach suggested in the current proposal is the multidimensional multi-criteria modeling and analysis involved with it. The application of the latest techniques involves economic, technical, and social aspects. Moreover, each of the innovative solutions depends upon a wide array of variables including weather conditions, terrain, rainfall data, road networks, and many others.

2.3 Water Harvesting and Collection

Access to clean water is essential for health. Unfortunately, many regions in the world do not have access to sufficient clean water. Consequently, many diseases have been reported due to this issue, including cholera, typhoid fever, and other waterborne illnesses [34].

Several water harvesting methods have been adopted to meet the need for clean water. Rainwater harvesting is commonly applied in regions that have enough rainfall [35]. This method is reported to have positive economic and environmental impacts in a developing country like Kenya [36]. Another important aspect of this system is the assurance of food security with the increase in irrigable lands [37].

However, rainwater harvesting systems need integration with storage containers and heavily depend upon the local environment [38]. Specific to arid regions, the collection of collector-drainage water is highly crucial due to the scarcity of rainfall in them. Apart from collection, sufficient processing of the water is also important for the use of collected water for irrigation purposes [39].

Permeable pavement is considered a vital part of a sustainable drainage system. On the other hand, it also aids the utilization of rainwater. The construction of permeable pavement consists of a drainage layer and a bedding system over the base layers to allow the passage of water [40]. Some of the advantages of these pavements are in reducing the impacts of stormwater by reducing frequency, volume, and duration. These pavements have also been reported to improve the quality of water by reducing temperature, pollutant concentrations, and pollutant loadings of suspended solids, heavy metals, polyaromatic hydrocarbons, and some nutrients [41]. Other benefits of permeable pavements include the mitigation of noise and the promotion of evaporative cooling [42]. Having said that, much consideration is required to be given to the suitability of permeable pavements for low-speed urban roads. The issues of structural stability, hydrologic performance, water quality, and maintenance are considered to be crucial [43]. Other points of concern regarding permeable pavements include scientific mechanisms for their evaluation and feasibility and guidelines for construction [42].

Among other water harvesting methods, dew and fog harvesting is also commonly found in the literature [44]. Wastewater harvesting is another method that could be used for groundwater maintenance or recreational activities [35]. Another method of water harvesting involves the use of atmospheric water vapor. This can be done using heat exchangers in the form of hot and cold surfaces. Such a system is found to be more energy efficient and economically feasible than other methods [45]. But the potential benefits of this system depend upon the energy source used. Some of the previous studies have utilized solar panels in conjunction with atmospheric water harvesting systems and found them to be very efficient [46]. Other improvements in this system include integrated use of a water treatment component so that the harvested water can fulfill the demands of a household without any health issues [47].

All the abovementioned methods require considerable cost and resources for their establishment and implementation on a wider scale. This points to the need for optimization of the planning and design of these systems to efficiently utilize the available national budget, land, and resources. In this regard, water harvesting through condensation of air-conditioning units is considered to be an economically viable option. Research suggests that the cost of such a method is only incurred when water is used for drinking purposes [48]. However, a more comprehensive study shows higher benefits of rainwater harvesting, compared to air-conditioning condensate harvesting, in terms of life cycle costs. This study, done by [49], uses an evidence-based approach with data from several projects to show the benefits of rainwater harvesting, which are at least 45% more than its counterpart.

2.4 Computational Methods

The current computation methods use artificial intelligence, with a growing use of deep learning approaches to handle such problems. However, it has been primarily used for forecasting and classification with little or no application in the field of optimization and decision-making [50]. This is linked with the lack of explanatory power often attached to artificial intelligence techniques. If the same is applied to optimization, then it would require formulation of the problem using an operations research methodology involving multiple combinations of best-case scenarios. Such an application is not available in the current literature. Moreover, other approaches used for decision-making (such as fuzzy logic systems) do not work well with large datasets.

2.5 Concluding Remarks

Table 1 summarizes the literature, clearly showing the importance of water management and water quality monitoring for a safe and healthy society. It also highlights the need for more research and development efforts to improve relevant systems. Smart real-time systems are deemed to be more efficient and effective in ensuring the quality of water, reducing energy consumption and other environmental impacts, and optimizing the resources.

IoT-based systems have been highly advocated for their real-time monitoring and data processing, which can be handled by a controller equipped with an accurate and efficient modeling mechanism. Some of the challenges indicated in this regard are the improvement of accuracy, placement and networking of sensors, handling vast amounts of data of varying nature, inclusion of necessary spatial and temporal details, and need to process it in a timely manner.

Table 1 Summary of literature

Topic	Summary
Water system monitoring	Monitoring water quality is a global priority linked to WHO standards and MDGs. High prices and scarcity are problems in arid areas. Real-time monitoring has been improved by SCADA and IoT systems, and proactive management is made possible by smart technologies like sensors and controllers. However, infrastructure, rather than water quality, is the primary focus of IoT applications. For quality monitoring, wireless sensors and AI-based models have been employed; in spite of that, issues with sensor location and model accuracy still exist
Water demand prediction	Reactive monitoring and water meter readings are the foundation of current systems. Accurate forecasts are hampered by constraints in temporal and spatial data. There is underutilization of new technologies (such as permeable pavements and rainwater collecting), while desalination is popular yet expensive. Integrating technical, economic, and social elements with a range of influencing factors is necessary for a thorough approach
Water harvesting and collection	Rainwater, fog/dew, wastewater, and atmospheric water harvesting are some of the techniques. Although rainwater systems have beneficial effects on the environment and the economy, local variables must be taken into account. Permeable pavements reduce the effects of rainfall and facilitate the use of rainwater. Air-conditioning condensate offers a viable source, but rainwater harvesting has superior life cycle benefits. All methods require optimization for cost-effectiveness
Computational methods	AI and deep learning are mainly used for forecasting and classification. Optimization and decision-making applications are limited due to lack of interpretability. Operational research and fuzzy logic approaches are underexplored for large datasets, highlighting a research gap

In spite of the looming danger of scarcity of clean water and the health problems associated with it, which are already materializing in some regions, the use of innovative methods of water harvesting is almost nonexistent in most of the countries in the world. Limitation of resources and methods to optimize their planning and design could be a major hurdle in the advancement of these methods and their use at a wider scale. Among these methods, rain harvesting is found to be the most common and beneficial in various ways. However, further research is required to enhance the efficiency of this system. Permeable pavements can be used to support the rainwater harvesting process, but their use in specific conditions requires due consideration due to a lack of evaluation mechanisms and construction guidelines.

The proposed framework of this research provides the necessary integration between different aspects of water quality and demand management. It also adopts diversification in handling the demand aspects. More importantly, it focuses on implementing the computational powers of quantum computing for uplifting the currently available models and control systems.

3 Digital Twin

Digital twin technology has recently gained attention for urban water management. A digital twin is a virtual replica of a physical system that updates in real time with data from sensors, allowing testing of strategies before implementation, and supports adaptive control [51].

Some of the relevant applications of digital twin models include monitoring of distribution networks, integration of desalination plants, and simulation of demand growth [52]. These models can combine hydraulic models, energy models, and socioeconomic variables, making them suitable for urban settings. Harnessing the abilities of edge cloud computing, they can process IoT data at scale and provide real-time decision support [53].

This chapter builds on these developments by embedding digital twin technology into a broader optimization framework.

3.1 AI and Hybrid Optimization Approaches

AI methods, which constitute the core of digital twin models, are increasingly used to complement traditional optimization. Machine learning models improve demand forecasting [54], leakage detection [55], and water quality prediction [56]. Reinforcement learning (RL) has been applied to optimize pumping schedules and reduce energy costs [57].

In addition, some studies have explored AI techniques for water management-related fields with successful implementation. Some examples of these efforts are mentioned below:

- Graph neural networks (GNNs) for analyzing flow in complex water networks [58]
- Transformer models for long-range demand forecasting [59]
- Metaheuristic algorithms such as genetic algorithms (GAs) and particle swarm optimization (PSO) for allocation and scheduling [60]
- Mixed-integer linear programming (MILP) for facility location and resource distribution [61]
- Stochastic optimization for handling uncertainty in rainfall and demand [62]

These methods are practical and deployable. They provide near-optimal solutions for large-scale, real-world problems, making them suitable for Bahrain's urban water challenges.

3.2 Quantum Computing as a Future Direction

Quantum computing has been proposed as an alternative for optimization problems that are computationally intensive. Especially for NP-hard and NP-complete optimization, factorization, and unstructured search, application of quantum algorithms has truly been found advantageous [63].

In water management, optimization problems such as network flow, desalination scheduling, and multi-objective trade-offs share features with these classes. Algorithms like the quantum approximate optimization algorithm (QAOA) are customized solutions to aid in such tasks [64].

However, current quantum hardware and algorithms are not yet mature enough for deployment. For this reason, quantum computing is not the core of this chapter. Instead, it is discussed herein as a possible extension for future research. A quantum-ready framework would require reformulation of optimization problems, and that work would be better suited to a dedicated study.

4 Proposed Framework

The proposed framework is shown in Fig. 1. At the core of the framework is the water security issue. The central integrating strategy leverages quantum computing's superior processing power for optimization, simulation, and forecasting. The digital twin model can dramatically enhance the capacity to process large datasets and run complex models far beyond the capabilities of classical computers, making it ideal for multifaceted challenges like water management.

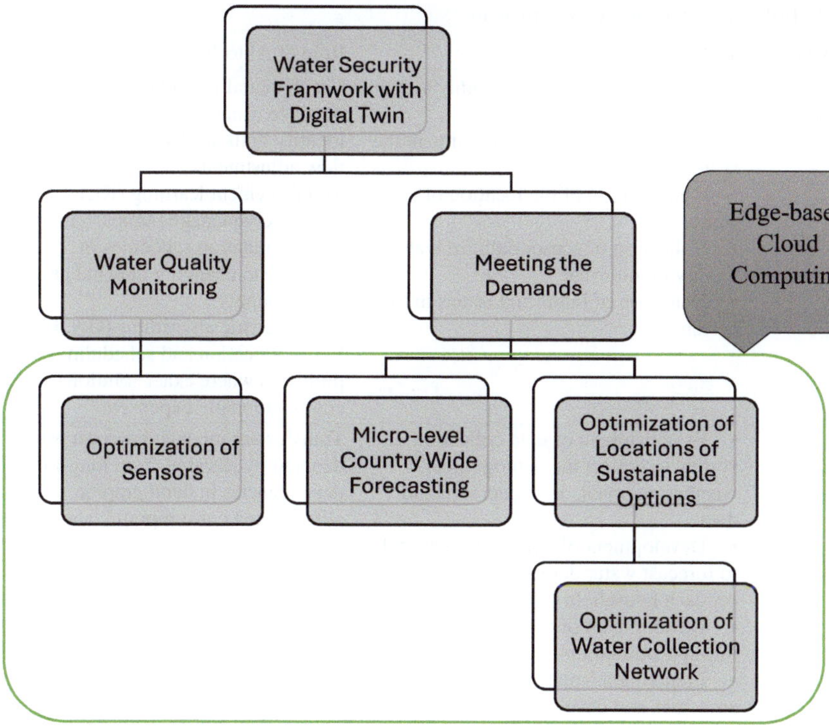

Fig. 1 Proposed framework

4.1 Key Components of the Framework

As stated above, the proposed framework adopts a multifaceted and comprehensive approach to mitigate water scarcity issues. The implementation of this framework deals with various approaches to water scarcity management, namely political ecology of water scarcity, water insecurity, and water economics. The tasks under each component are presented in Table 2. The role of AI computational technologies computing for each case is also provided in this table.

The optimization of sensor locations, locations of water resources, treatment plants, and water collection systems will deal with the political ecology and water economics. The optimized solutions will maximize the economic benefits within the available resources and constraints. It will also ensure that the installed facilities and development plans are based on technical aspects with minimum impact of nontechnical issues. Water demand prediction at the microlevel covers political ecology and water insecurity. It will cater to the human demand aspect, and the microlevel prediction may indirectly identify the negative trends of human consumption. It will also aid in informed decision-making and devising political vision to deal with water security issues in the long run.

Table 2 Tasks under the proposed framework

Component	Tasks	Role of AI technologies
Water quality monitoring	• Development of water supply system inventory • Identification of critical points in the system • Optimization of the location of sensors • Collection of sensor data for water quality monitoring • Detection of faults and issues in the system	Graph neural networks (GNNs): Analyze water distribution networks, identify bottlenecks, and recommend flow adjustments Reinforcement learning (RL): Adaptively manage pumps, valves, and desalination schedules in response to changing demand and energy supply Metaheuristic algorithms (GA, PSO): Solve allocation and scheduling problems where exact solutions are computationally expensive
Microlevel demand forecasting	• Establishment of data collection system to collect the demographic, weather, regional, and water demand data • Development of a prediction model to forecast water demand on a microlevel for each household • Development of future scenarios through simulation • Preparing adaptable and flexible strategies and development plans for each possible scenario	Transformer models: Forecast water demand by learning from long-range dependencies in demographic, climate, and consumption data
Optimization of location of water supply options	• Review of possible traditional and innovative water supply systems that could be employed in the local conditions • Collecting the set of variables affecting the water generation through each method • Few relevant examples: aqua harvesting, desalination • Developing a geographic information system (GIS) incorporating the spatial data with the required variables and systems identified above • Developing quantum computing-based optimization algorithms for each of the supply systems • Integrating the optimization system with GIS for real-time planning • Feasibility studies of optimized systems • Optimizing the investments in the mix of optimized systems to satisfy the resource and budget constraints while addressing the required demand	Metaheuristic algorithms (genetic algorithm, particle swarm optimization): Solve allocation and scheduling problems where exact solutions are computationally expensive Mixed-integer linear programming (MILP): Optimize facility location, storage capacity, and plant operations

(continued)

Table 2 (continued)

Component	Tasks	Role of AI technologies
Optimization of location of water collection systems	• Collecting data on the rainfall and drainage patterns in the region • Integrating the data into GIS • Identification of water collection techniques suitable for the region • Possibilities: permeable pavements, water reservoirs, stormwater drainage systems, etc. • Developing optimization tools based on quantum computing, for the location of water collection systems • Integrating the optimization tools with GIS • Acquiring future land use and development plans • Integrating land use and development patterns in GIS • Optimizing the location of water collection systems • Optimizing the investment plans based on future land use and development plans	Mixed-integer linear programming (MILP): Optimize facility location, storage capacity, and plant operations Stochastic optimization: Incorporate uncertainty from rainfall variability, demand fluctuations, and energy prices

4.1.1 Water Quality Monitoring

Real-time data processing from numerous water quality sensors located throughout the nation can be improved by quantum computing. Applied to Bahrain's distribution network, GNNs can identify pressure points, highlight leakage risks, and suggest flow adjustments.

4.1.2 Meeting the Demands

It is difficult to balance the supply and demand for water, particularly in light of urbanization and climate change. Real-time supply-demand dynamics can be simulated and managed via the combination of predictive power of transformer models and simulation power of RL, which accounts for factors like industrial demand, seasonal usage, and population increase.

4.1.3 Optimization of Sensors

Widespread, trustworthy sensor networks are essential for effective water security. In order to minimize redundancy and maximize data coverage and accuracy, metaheuristic algorithms and computing techniques can assist in determining the best location and calibration of sensors throughout Bahrain.

4.1.4 Microlevel Countrywide Forecasting

Rainfall, groundwater recharge, evaporation rates, and other hydrological parameters can all be precisely predicted at the microlevel using forecasting models enhanced with transformer models for prediction and stochastic optimization for incorporation of meteorological and hydrological uncertainty. This makes it possible to plan water resources and respond to emergencies with greater initiative and knowledge.

4.1.5 Optimization of Locations of Sustainable Options

For optimal benefit, sustainable water choices like desalination plants, rainwater harvesting, and graywater reuse systems must be positioned carefully. To choose the best sites for these interventions, multivariable spatial optimization models based on metaheuristic algorithms, including mixed-integer linear programming (MILP), GA, and PSO, can examine demand, infrastructure, climate, and topography.

4.1.6 Optimization of Water Collection Network

By employing metaheuristic algorithms to simulate different network topologies, Bahrain's water collection and distribution network can become more efficient. These simulations assist in determining the most economical and energy-efficient methods for gathering and delivering water by minimizing losses and enhancing delivery.

4.2 Data Acquisition

A crucial component of every professional or research project is the collection of data of acceptable precision and accuracy levels. Planning and implementation of a successful strategy require accurate data in sufficient quantity to make informed decisions and forecast future scenarios. The water supply data would involve long-term consumption and rainfall data. Such dataset can be taken from World Bank databases or local ministries and departments (such as the Electricity and Water Authority (EWA) for Bahrain). Moreover, the comprehensive management of this issue would require the integration of this data with the terrain, road network, and pipeline network geographical data. In the context of Bahrain, as-built and development drawings from the Survey and Land Registration Bureau (SLRB), the Ministry of Works (MoW), and the EWA can be acquired to achieve this integration. The entire system is proposed to operate on an edge cloud computing architecture to enable distributed processing of sensor data and real-time decision-making.

4.3 Strategic Implications for Water-Scarce Countries

This framework is expected to have global generic application due to the well-established use of state-of-the-art technologies. However, it is particularly critical for Bahrain and other countries with arid climate due to the following reasons.

In the absence of sufficient rainfall, most of these countries rely on desalination for fulfilling the consumption needs of their population. This process is energy intensive and requires a considerable amount of time and resources for its installation. Alternative approaches to demand fulfillment will ensure water security and energy efficiency and incur numerous economic benefits.

Arid regions are highly vulnerable to climate change; this is due to the lack of green areas coupled with the increase in built-up space and energy consumption for livelihood and transportation. This situation leads to reduced groundwater recharge and unpredictable rainfall. Retaining and maximizing the use of rainwater will help to transform the overall climatic conditions with the enhancement of green spaces within the developed areas.

Population growth and quantum computing are phenomena which have become popular globally due to their far-reaching and multifaceted impacts, in their own respective ways. The growing urban population increases pressure on existing infrastructure, rendering it outdated and incapacitated. The massive investments required to frequently upgrade and enhance infrastructure are infeasible for any government. The proposed framework provides a smart and efficient utilization of existing infrastructure with the incorporation of advanced monitoring techniques, which provide the necessary input for high-performance predictive computing. The proposed framework also proposes subtle changes in the existing construction techniques, such as permeable pavements, which would aid in achieving the maximum benefits from the proposed methodology.

The proposed framework, if adopted by any country, provides an added advantage for the participant country to become a world leader by integrating a digital twin model into the national water strategy. Countries with such futuristic and advanced strategies can lead the region with the deployment of advanced technologies for sustainability. Furthermore, it opens avenues for partnerships with global tech firms and research institutions, positioning the country as a smart water innovation hub.

4.4 Alignment with Sustainable Development Goals (SDGs)

The end of the last century experienced a surge in environmental degradation and social insecurities. The persistence of this issue was recognized by global leaders who grouped together under the umbrella of the United Nations to agree to and sign the Sustainable Development Goals (SDGs)—2030 [65]. The combination of 17 SDGs was signed by 193 member countries, which came into effect from 2016 onwards [66]. The active pursuance of the SDGs and their eventual achievement are considered critical for ensuring global environmental sustainability and social security. To this end, it is vital that essential changes should be made in governance and financial practices with the necessary support of external factors [67]. The core of these SDGs revolves around the development of healthy societies and is aligned with collective well-being. This necessitates addressing all the sectors related to the safe and secure provision of basic utilities [68].

The proposed framework is developed with due consideration to accelerating and facilitating the achievement of multiple SDGs. It focuses on SDG 6 (Clean Water and Sanitation) directly as it deals with water security issues from multiple aspects. It also deals with SDG 9 (Industry, Innovation, and Infrastructure) as it involves using innovative solutions and computing technology for optimizing the infrastructure development and monitoring related to water supply. Moreover, the issue of water security impacts the entire community and its sustainability; hence, SDG 11 (Sustainable Cities and Communities) is also linked with this project. Lastly, the project would also identify patterns of water consumption and management of the water supply network, which is under the domain of SDG 12 (Responsible Consumption and Production).

5 Conclusion

This research aims to address the issue of water scarcity in arid countries. It proposes the use of high-performance computational power of quantum mechanics with the integration of maximizing utilization of current infrastructure development and construction techniques.

The proposed framework addresses the issue of water security by minimizing water losses, optimizing the use of current water supply and management practices, and maximizing the retention and use of natural water available through whatever amount of rainfall is available. A high-performance digital twin model incorporating various computing technologies would be incorporated into the prediction of demand, optimization of water supply projects, and monitoring systems. The deployment of the digital twin is expected to provide the essential dimensions of efficiency, accuracy, and time saving.

The proposal of the current research is based on the proven and reported trends of water management, advantages of modern practices, and AI-based computing. Hence, it is expected to be valid and reliable. Moreover, it could be considered equally beneficial for other regions having arid regions, such as sub-Saharan and the South Asian subcontinent. Countries in these regions have the common factor of arid climate and/or being in a river basin with the countries in the Middle East. However, the current status of the framework is purely theoretical, and the next step towards development and deployment of such a system is simulation and prototype testing. Urban design and planning tools could be utilized for pursuing future research in this direction. A key issue raised in the literature is the role of quantum computing. Developing a quantum-ready solution would require reformulation of optimization problems, which goes beyond the scope of this research. Therefore, quantum computing is recommended as a possible extension for future research.

Funding Acknowledgment: This work was supported by the Deanship of Scientific Research, Vice Presidency for Graduate Studies and Scientific Research, King Faisal University, Saudi Arabia [Grant No.: KFU 261393].

AI Use Disclosure Statement The authors used generative AI technologies only for proofreading. This includes minor corrections to grammar, spelling, and syntax. No AI was used for ideation, data analysis, or writing original content.

References

1. Falkenmark M (2013) Growing water scarcity in agriculture: future challenge to global water security. Philos Trans R Soc A Math Phys Eng Sci 371(2002):20120410
2. Wang J, Li Y, Huang J, Yan T, Sun T (2017) Growing water scarcity, food security and government responses in China. Glob Food Sec 14:9–17
3. Eckstein GE (2009) Water scarcity, conflict, and security in a climate change world: challenges and opportunities for international law and policy. Wis Int'l LJ 27:409
4. Rosengrant MW, Cai X (2001) Water scarcity and food security: alternative futures for the 21st century. Water Sci Technol 43(4):61–70
5. Fereres E, Orgaz F, Gonzalez-Dugo V (2011) Reflections on food security under water scarcity. J Exp Bot 62(12):4079–4086
6. Nechifor V, Winning M (2018) Global economic and food security impacts of demand-driven water scarcity—alternative water management options for a thirsty world. Water (Basel) 10(10):1442
7. Schmidt GA et al (2017) Practice and philosophy of climate model tuning across six U.S. modeling centers. Geosci Model Dev 10:3207–3223. https://doi.org/10.5194/gmd-10-3207-2017
8. Wutich A, Beresford M, Montoya T, Radonic L, Workman C (2022) Water security and scarcity. In: Oxford research encyclopedia of anthropology
9. Qureshi AS (2020) Challenges and prospects of using treated wastewater to manage water scarcity crises in the Gulf Cooperation Council (GCC) countries. Water (Basel) 12(7):1971
10. El-Sayed MK, Mansour RS (2017) Water scarcity as a non-traditional threat to security in the Middle East. India Q 73(2):227–240
11. Mfarrej MFB (2025) Exploring the nexus between climate change, water scarcity, and security dynamics in the Middle East and North Africa. Next Res 2025:100168
12. Aisopou A, Stoianov I, Graham NJ (2012) In-pipe water quality monitoring in water supply systems under steady and unsteady state flow conditions: a quantitative assessment. Water Res 46(1):235–246
13. Bartram J et al (2014) Global monitoring of water supply and sanitation: history, methods and future challenges. Int J Environ Res Public Health 11(8):8137–8165
14. Perveen S (2023) Drinking water quality monitoring, assessment and management in Pakistan: a review. Heliyon 9(3)
15. Bandurin MA, Yurchenko IF, Bandurina IP, Bandurin AP (2021) Problems and prospects of using mobile devices when performing operational monitoring of water supply structures. IOP Conf Ser Mater Sci Eng 1111(1):012006
16. Haider H, Sadiq R, Tesfamariam S (2014) Performance indicators for small- and medium-sized water supply systems: a review. Environ Rev 22(1):1–40

17. Bathre M, Das PK (2022) Water supply monitoring system with self-powered LoRa based wireless sensor system powered by solar and hydroelectric energy harvester. Comput Stand Interfaces 82:103630
18. Dobriceanu M, Bitoleanu A, Popescu M, Enache S, Subtirelu E (2008) SCADA system for monitoring water supply networks. WSEAS Trans Syst 1(7):1070–1079
19. Yasin HM et al (2021) IoT and ICT based smart water management, monitoring and controlling system: a review. Asian J Res Comput Sci 8(2):42–56
20. Pérez-Padillo J, García Morillo J, Ramirez-Faz J, Torres Roldán M, Montesinos P (2020) Design and implementation of a pressure monitoring system based on IoT for water supply networks. Sensors 20(15):4247
21. Roy A, Mukhopadhyay S, Roy S (2022) IoT based water quality monitoring system. In: 2022 second international conference on computer science, engineering and applications (ICCSEA). IEEE, pp 1–4
22. Kim Y, Schmid T, Charbiwala ZM, Friedman J, Srivastava MB (2008) NAWMS: nonintrusive autonomous water monitoring system. In: Proceedings of the 6th ACM conference on Embedded network sensor systems, pp 309–322
23. Robles-Durazno A, Moradpoor N, McWhinnie J, Russell G (2018) A supervised energy monitoring-based machine learning approach for anomaly detection in a clean water supply system. In: 2018 international conference on cyber security and protection of digital services (Cyber Security). IEEE, pp 1–8
24. Zulkifli CZ et al (2022) IoT-based water monitoring systems: a systematic review. Water (Basel) 14(22):3621
25. Stoianov I, Nachman L, Whittle AJ, Madden S, Kling R (2006) Sensor networks for monitoring water supply and sewer systems. In: Proc. 8th Intl. symposium on water distribution systems analysis, (WDSA2006), Cincinnati
26. Misiunas D (2005) Failure monitoring and asset condition assessment in water supply systems (monograph). Division for Industrial Electrical Engineering and Automation. Department of Industrial Electrical Engineering and Automation, Lund Institute of Technology. https://lucris.lub.lu.se/ws/portalfiles/portal/5704571/26607.pdf
27. Liu J, Wu D, Mohammed H, Seidu R (2024) A novel method for anomaly detection and signal calibration in water quality monitoring of an urban water supply system. Water (Basel) 16(9):1238
28. Jabbar WA et al (2024) Development of LoRaWAN-based IoT system for water quality monitoring in rural areas. Expert Syst Appl 242:122862
29. Bharani Baanu B, Jinesh Babu KS (2022) Smart water grid: a review and a suggestion for water quality monitoring. Water Supply 22(2):1434–1444
30. B. Rousso, S. Brady, and J. J. Gong, "Effectiveness of rainwater tank-to-toilet scheme in the Quay estate, Torquay, Victoria," 2024
31. Díaz S, González J (2021) Temporal scale effect analysis for water supply systems monitoring based on a microcomponent stochastic demand model. J Water Resour Plan Manag 147(5):4021023
32. Kumar P, Choudhury D (2024) Innovative technologies for effective water resources management. In: Water crises and sustainable management in the global south, pp 555–594
33. Marzooq M, Alsabbagh M, Al-Zubari W (2018) Energy consumption in the municipal water supply sector in the Kingdom of Bahrain. Computat Water Energy Environ Eng 7(03):95
34. Jarimi H, Powell R, Riffat S (2020) Review of sustainable methods for atmospheric water harvesting. Int J Low-Carbon Technol 15(2):253–276
35. Zang J, Kumar M, Werner D (2021) Real-world sustainability analysis of an innovative decentralized water system with rainwater harvesting and wastewater reclamation. J Environ Manag 280:111639
36. Malesu MM (2006) Rainwater harvesting innovations in response to water scarcity: the Lare experience (No. 5). World Agroforestry Centre

37. Komariah K, Senge (2013) The development of water harvesting research for agriculture. Rev Agric Sci 1. https://doi.org/10.7831/ras.1.31
38. Chen K, Tao Y, Shi W (2022) Recent advances in water harvesting: a review of materials, devices and applications. Sustainability 14(10):6244
39. Sherov A, Urinboev S (2020) Innovative technologies in the effective use of water resources. IOP Conf Ser Mater Sci Eng 883(1):012144
40. Scholz M, Grabowiecki P (2007) Review of permeable pavement systems. Build Environ 42(11):3830–3836
41. Drake JA, Bradford A, Marsalek J (2013) Review of environmental performance of permeable pavement systems: state of the knowledge. Water Qual Res J Can 48(3):203–222
42. Zhou J et al (2024) Review of the development and research of permeable pavements. Hydrol Process 38(6):e15179
43. Weiss PT, Kayhanian M, Gulliver JS, Khazanovich L (2019) Permeable pavement in northern North American urban areas: research review and knowledge gaps. Int J Pavement Eng 20(2):143–162
44. Pirouz B, Palermo SA, Turco M (2021) Improving the efficiency of green roofs using atmospheric water harvesting systems (an innovative design). Water (Basel) 13(4):546
45. Sanaye S, Shourabi A, Borzuei D (2023) Sustainable water production with an innovative thermoelectric-based atmospheric water harvesting system. Energy Rep 10:1339–1355
46. Xiang T, Xie S, Chen G, Zhang C, Guo Z (2025) Recent advances in atmospheric water harvesting technology and its development. Mater Horiz 12(4):1084–1105
47. Ferwati MS (2019) Water harvesting cube. SN Appl Sci 1(7):779
48. Elbrashy A, Vafai K, Elshennawy A, Ayman M, Elgebaly A, Rashad M (2025) Harvesting of condensate water from air conditioners in large institutions as a sustainable resource. Water Resour Manag:1–24
49. Ghimire SR, Johnston JM, Garland J, Edelen A, Ma XC, Jahne M (2019) Life cycle assessment of a rainwater harvesting system compared with an AC condensate harvesting system. Resour Conserv Recycl 146:536–548
50. Benti NE, Chaka MD, Semie AG (2023) Forecasting renewable energy generation with machine learning and deep learning: current advances and future prospects. Sustainability 15(9):7087. https://doi.org/10.3390/su15097087
51. Chaudhari P, Gangane C, Lahe A (2021) Digital twin in industry 4.0 a real-time virtual replica of objects improves digital health monitoring system. In: International conference on information systems and management science. Springer International Publishing, Cham, pp 506–517
52. Henriksen HJ et al (2022) A new digital twin for climate change adaptation, water management, and disaster risk reduction (HIP digital twin). Water (Basel) 15(1):25
53. Wang Y, Fang J, Cheng Y, She H, Guo Y, Zheng G (2023) Cooperative end-edge-cloud computing and resource allocation for digital twin enabled 6G industrial IoT. IEEE J Sel Top Signal Process 18(1):124–137
54. Narayanan LK, Sankaranarayanan S, Rodrigues JJ, Kozlov S (2020) Water demand forecasting using deep learning in IoT enabled water distribution network: DL-water demand forecasting for water distribution design. Int J Comput Commun Control 15(6)
55. X. Fan, X. Zhang, and . X Yu, "Machine learning model and strategy for fast and accurate detection of leaks in water supply network," J Infrastruct Preserv Resil, vol. 2, no. 1, 10, 2021
56. Dogo EM, Nwulu NI, Twala B, Aigbavboa C (2019) A survey of machine learning methods applied to anomaly detection on drinking-water quality data. Urban Water J 16(3):235–248
57. Pei S, Hoang L, Fu G, Butler D (2025) Real-time pump scheduling in water distribution networks using deep reinforcement learning. J Water Resour Plan Manag 151(6):4025012
58. Kerimov B et al (2023) Assessing the performances and transferability of graph neural network metamodels for water distribution systems. J Hydroinf 25(6):2223–2234
59. Grigsby J, Wang Z, Nguyen N, Qi Y (2021) Long-range transformers for dynamic spatiotemporal forecasting. arXiv

60. Agarwal M, Srivastava GMS (2018) Genetic algorithm-enabled particle swarm optimization (PSOGA)-based task scheduling in cloud computing environment. Int J Inf Technol Decis Mak 17(04):1237–1267
61. Yang Y, Zhang S, Xiao Y (2015) An MILP (mixed integer linear programming) model for optimal design of district-scale distributed energy resource systems. Energy 90:1901–1915. https://doi.org/10.1016/j.energy.2015.07.013
62. Santos AMP, Fagerholt K, Soares CG (2023) A stochastic optimization algorithm for the supply vessel planning problem under uncertain demand and uncertain weather conditions. Ocean Eng 278:114428
63. Katsifas FM, Moustakidis C, Konofaos N (2021) Quantum algorithms for solving combinatorial optimization problems. Aristotle University of Thessaloniki
64. Urgelles H, Picazo-Martinez P, Garcia-Roger D, Monserrat JF (2022) Multi-objective routing optimization for 6G communication networks using a quantum approximate optimization algorithm. Sensors 22(19):7570
65. Pedersen CS (2018) The UN sustainable development goals (SDGs) are a great gift to business! Procedia CIRP 69:21–24
66. Mori Junior R, Fien J, Horne R (2019) Implementing the UN SDGs in universities: challenges, opportunities, and lessons learned. Sustain J Record 12(2):129–133
67. Pizzi S, Caputo A, Corvino A, Venturelli A (2020) Management research and the UN sustainable development goals (SDGs): a bibliometric investigation and systematic review. J Clean Prod 276:124033
68. Saxena A, Ramaswamy M, Beale J, Marciniuk D, Smith P (2021) Striving for the United Nations (UN) sustainable development goals (SDGs): what will it take? Discov Sustain 2(1):20

Attitude Control of CubeSats: Design and Performance Comparison of Similar Proportional Derivative and Backstepping Techniques

Halima Boussadia, Abdelkrim Meche, Nabil Boughanmi, and Arezki Mohammed Si Mohammed

Abstract With the advancement of aerospace technology, the issue of spacecraft attitude control has grown in significance, especially in light of CubeSats, which provide inexpensive access to space with a variety of mission capabilities. For satellite payloads like antennas, solar panels, and observation tools to be oriented correctly, the Attitude and Orbit Control System (AOCS), which can use active control methods or passive stabilization procedures, is responsible for this duty. Because of their resilience and efficiency in handling external disturbances and nonlinear satellite dynamics, proportional-derivative (PD) and backstepping controllers are among the many nonlinear control techniques that have been developed in the literature. The difficulty of attitude control for the 12U CubeSat 1HOPSAT (first-generation high optical performance satellite) is addressed in this work. Based on Lyapunov stability theory, two distinct control strategies, a backstepping controller and a similar proportional-derivative controller, are developed and examined. Their accuracy, stability, and convergence time are highlighted in the comparison analysis, which offers helpful guidance for choosing effective control schemes for CubeSat missions.

H. Boussadia (✉)
Department of Automatic, Laboratory of Signals and Systems Data, USTO-MB University, Oran, Algeria
e-mail: halima.boussadia@univ-usto.dz

A. Meche
Department of Automatic, Laboratory of Signals and Images, USTO-MB University, Oran, Algeria
e-mail: abdelkrim.meche@univ-usto.dz

N. Boughanmi
Department of Electronic, Laboratory of Signals and Systems Data, USTO-MB University, Oran, Algeria
e-mail: nabil.boughanmi@univ-usto.dz

A. M. Si Mohammed
Department of Space Mechanic Research, Centre of Satellite Development, Oran, Algeria
e-mail: masimohammed@asal.cds.dz

M. Mirrashid et al. (eds.), *Optimization and Adaptive Control Algorithms*, Emerging Trends in Mechatronics, https://doi.org/10.1007/978-981-95-8387-4_3

Keywords CubeSat · Attitude control · Backstepping control · Similar proportional-derivative control

1 Introduction

The satellite's objective is to keep it under control when it is in orbit. We need to know the satellite's attitude in order to complete this operation [1]. The spacecraft Attitude and Orbit Control System (AOCS) is primarily responsible for this latter task [2]. Therefore, spacecraft attitude control is an important aspect of spacecraft research.

Due to advancements in aerospace technology, spacecraft attitude control has garnered significant attention in recent years [3]. The satellite needs to be stabilized at a particular attitude for a space mission to be successful [4]. This stabilization is a fundamental operation that follows the launch and deployment phases of the satellite, and it is achieved through onboard corrections by the Attitude and Orbit Control System (AOCS). Either passive method, by taking advantage of basic properties (gravitational, magnetic, etc.), or active method, by employing a controller that determines the required control torques to change the satellite's attitude to a desired orientation, this involves understanding the dynamic model of the satellite and controlling its orientation to ensure that antennas, solar panels, or observation instruments are pointed in the correct direction.

Numerous nonlinear controls have been brought to light in satellite attitude control. PID control [5], PD control [6], sliding mode control [7], backstepping technique [8–10], and adaptive control [11] are some examples of these controllers.

The advent of tiny satellites, or CubeSats, is greatly influencing space technology development, especially in the nanosatellite class. Because of their more straightforward designs, quicker development times, and far lower launch prices than conventional satellites, CubeSats are becoming more and more popular. Because of these qualities, CubeSats are becoming more and more appealing for a range of uses.

CubeSats are being used for a variety of on-orbit missions, such as space exploration [12, 13], telecommunications constellations [14, 15], and Earth observation [16]. Because of their adaptability and affordability, space has become more accessible to researchers, academic institutions, and private companies, enabling them to take part in space missions that were previously unaffordable. To fully utilize the potential of the sophisticated payloads that CubeSats may transport, a number of issues must be resolved despite their potential. The need for advanced payloads on board CubeSats is increasing due to their compact size and economic considerations. The creation of increasingly powerful instruments and sensors appropriate for space applications is made possible by developments in the electrical and aerospace industries, which are the main drivers of this demand.

The field of CubeSat attitude control has seen substantial advancements through diverse research endeavors aiming to achieve optimal performance, particularly for nonlinear systems. Various control strategies have been explored, each offering unique advantages in addressing the challenges inherent to satellite dynamics. Proportional-derivative (PD) control combines feedforward and feedback control by operating on both the current and predicted process conditions and, along with its variants, stands out as one of the most widely adopted methods for CubeSat attitude control. Its popularity stems from its ability to effectively handle external disturbances. For instance, in [17], a PD magnetic controller was implemented for a 1U CubeSat. In [18], a stabilizing PD-like control law was employed for stabilizing the attitude of a 3U CubeSat pointing toward Earth.

Finding a Lyapunov function that ensures the system's performance is the second method of Lyapunov, also known as the direct method, which forms the basis of backstepping control. This approach is based on the idea of system energy, where a positive definite Lyapunov function (like energy) is used in place of the system's energy, and its derivative needs to be negative. The backstepping technique works in a step-by-step or recursive manner. To guarantee that the system converges to its equilibrium state, a virtual control is calculated at each stage. Lyapunov functions, which gradually stabilize each stage of synthesis, can be used to do this.

During the satellite's design phase, the main goal of the Attitude and Orbit Control System (AOCS) designer is to determine the best command strategy for the satellite's attitude based on accuracy, convergence time, or both, depending on the mission requirements. As a result, a number of queries come up, such as which control strategy is best. How ought one to make this decision? In order to make the control mechanism feasible for satellite applications, how may the aforementioned issues be resolved? The main objective of this study is defined by these questions. As previously stated, the space system community is paying close attention to reliable control strategies for CubeSats, with a focus on their effectiveness and performance. In order to manage the 12U CubeSat 1HOPSAT (first-generation high optical performance satellite), this work's objective is to develop and compare two distinct controllers: the similar proportional-derivative controller and the backstepping controller.

The structure of the chapter is as follows: The satellite's kinematic and dynamic models are shown in Sect. 2. The suggested control laws' design is presented in Sect. 3. Section 4 illustrates results of the simulation, and Sect. 5 concludes the chapter.

2 Spacecraft Attitude Model

The equation for the spacecraft's dynamics in the inertial frame is determined by Euler's equations of motion. It is presented as follows [19, 20]:

$$\boldsymbol{I}\dot{\omega}_s^I = \boldsymbol{C}_{ext} + \boldsymbol{I}\omega_s^I \times \left(\boldsymbol{I}\omega_s^I + \boldsymbol{h}\right) - \dot{\boldsymbol{h}} \quad (1)$$

where

$\boldsymbol{I} = [I_x \ I_y \ I_z]^T$ is the spacecraft moment of inertia.

$\boldsymbol{\omega}_s^I = \begin{bmatrix} \omega_x & \omega_y & \omega_z \end{bmatrix}^T$ is the vector of angular velocity in the inertial frame.

$\boldsymbol{h} = \begin{bmatrix} h_x & h_y & h_z \end{bmatrix}^T$ is the angular moment vector.

$\boldsymbol{C}_{ext} = \begin{bmatrix} C_x & C_y & C_z \end{bmatrix}^T$ is the external disturbance torque vector.

The kinematic equation is expressed as

$$\dot{\boldsymbol{q}} = \frac{1}{2}\Omega \boldsymbol{q} = \frac{1}{2}\Lambda(\boldsymbol{q}) \tag{2}$$

where

$\boldsymbol{q} = \begin{bmatrix} q_1 & q_2 & q_3 & q_4 \end{bmatrix}^T$ quaternion

$$\boldsymbol{\Omega} = \begin{bmatrix} 0 & \omega_{oz} & -\omega_{oy} & \omega_{ox} \\ -\omega_{oz} & 0 & \omega_{ox} & \omega_{oy} \\ \omega_{oy} & -\omega_{ox} & 0 & \omega_{oz} \\ -\omega_{ox} & -\omega_{oy} & -\omega_{oz} & 0 \end{bmatrix} \tag{3}$$

and

$$\Lambda(q) = \begin{bmatrix} q_4 & -q_3 & q_2 \\ q_3 & q_4 & -q_1 \\ -q_2 & q_1 & q_4 \\ -q_1 & -q_2 & -q_3 \end{bmatrix} \tag{4}$$

The transformation matrix $\boldsymbol{A}$ can be utilized for transforming the inertial reference body rates into the angular body rates referenced to the orbit coordinates [19].

$$\boldsymbol{\omega}_s^o = \boldsymbol{\omega}_s^I - \boldsymbol{A}\omega_o \tag{5}$$

where

$\boldsymbol{\omega}_s^o = \begin{bmatrix} \omega_{ox} & \omega_{oy} & \omega_{oz} \end{bmatrix}^T$ is the orbital angular velocity.

The attitude matrix that transforms any vector from the orbital frame to the body frame in terms of quaternions is presented as

$$A = \begin{bmatrix} q_1^2 - q_2^2 - q_3^2 + q_4^2 & 2(q_1q_2 + q_3q_4) & 2(q_1q_3 - q_2q_4) \\ 2(q_1q_2 - q_3q_4) & -q_1^2 + q_2^2 - q_3^2 + q_4^2 & 2(q_2q_3 + q_1q_4) \\ 2(q_1q_3 + q_2q_4) & 2(q_2q_3 - q_1q_4) & -q_1^2 - q_2^2 + q_3^2 + q_4^2 \end{bmatrix} \quad (6)$$

Since the quaternion lacks a clear physical interpretation, Euler angles can be used to determine the satellite's attitude. To transform a vector from the body coordinates to the reference orbital, the attitude transformation matrix $\boldsymbol{A}$ will be written as

$$A = \begin{bmatrix} c\psi c\theta + s\psi s\Phi s\theta & s\psi c\Phi & -c\psi s\theta + s\psi s\Phi c\theta \\ -s\psi c\theta + c\psi s\Phi s\theta & c\psi c\Phi & s\psi s\theta + c\psi s\Phi c\theta \\ c\Phi s\theta & -s\Phi & c\Phi c\theta \end{bmatrix} \quad (7)$$

where

θ: roll angle; φ: pitch angle; ψ: yaw angle.

The system's controllability has been demonstrated in [4].

3 Similar Proportional-Derivative and Backstepping Control Design

The theoretical underpinnings of two well-known control techniques, the backstepping control and the similar proportional-derivative (SPD) controller, are examined in this section. Because of their efficiency in controlling system dynamics and guaranteeing stability, both controllers are frequently utilized in a variety of control systems.

We start by investigating the SPD controller, a traditional control technique that uses derivative and proportional actions to accomplish the intended system performance. The backstepping controller, a more sophisticated control method created especially for nonlinear systems, will then be discussed.

3.1 *Similar Proportional-Derivative Controller*

A modified controller known as the similar proportional-derivative (SPD) controller is introduced in this section. Because it works with both the current and anticipated process conditions, the SPD control combines feedforward and feedback control. The error signal and its derivative are combined linearly to provide the control output in SPD control.

- Proportional control (P): This action is based on the system error. The proportional control action adjusts the control input in proportion to this error. The proportional gain K_p determines the sensitivity of the controller to the error.
- Derivative control (D): This action is based on the rate of change of the error, providing a predictive control action. It helps in damping the system's response and improving the stability by reacting to the error's trend. The derivative gain K_d determines the sensitivity of the controller to the rate of change of the error.

SPD control, as previously mentioned, relates the controller output to both the error and its derivative. This control method is mathematically expressed as

$$u(t) = K_p e(t) + K_d \frac{d}{dt} e(t) \tag{8}$$

where

$u(t)$ is the control input.
$e(t) = r(t) - y(t)$ is the error signal, with $r(t)$ being the desired point and $y(t)$ being the system output.
K_p is the proportional gain.
K_d is the derivative gain.

According to Eq. (8), the SPD controller is similar to a simplified PID controller with a zero integral term.

Based on the model developed in the second section, we design the control law for the maneuvering CubeSat.

We will briefly recall the dynamic and kinematic model of this CubeSat.

$$\dot{q}_0 = -\frac{1}{2} q_v^T \omega^0 \tag{9}$$

$$\dot{q}_v = \frac{1}{2}\left(q_0 I_{3\times3} + [q_v]_\times\right)\omega^0 \tag{10}$$

$$I_s \dot{\omega} = -\omega \times I_s \omega \tag{11}$$

The control law of SPD is proposed as

$$\tau_c = -K_p q_{ve} - K_d \omega^0 + \omega \times I_s \omega \tag{12}$$

where K_p and K_d are positive scalar constants.

The controller is developed to stabilize the satellite attitude to any desired quaternion vector q_{vd}, where the attitude error is defined by

$$q_{ve} = q_v - q_{vd} \tag{13}$$

Theorem 1 Let the dynamics of the satellite attitude be defined as in (9), (10), and (11). If the designed parameters K_p and K_d defined in (12) are positive, then the control input τ_c globally asymptotically stabilizes the attitude system to the equilibrium point $q_v = q_{vd}$, $\omega_0 = 0$ rad/s, $t \to \infty$.

Proof Thinking about the appropriate Lyapunov function:

$$V = \frac{1}{2}\left(\omega^0\right)^T I_s \omega^0 + 2K_p\left(1 - q_{0e}\right) \tag{14}$$

Considering that

$$1 - q_{0e} \geq \frac{1}{2}\left(1 - \left(q_{0e}\right)^2\right) = \frac{1}{2}\left(q_{ve}\right)^T q_{ve} \tag{15}$$

Calculating the derivative of (14) yields

$$\dot{V} = \left(\omega^0\right)^T I_s \dot{\omega}^0 \quad 2K_p \dot{q}_{0e} \tag{16}$$

Using Eqs. (9) and (11), the relative attitude dynamic error can be determined as

$$\dot{q}_{0e} = -\frac{1}{2}\left(q_{ve}\right)^T \omega^0 \tag{17}$$

and

$$\dot{q}_{ve} = \frac{1}{2}\left(q_{0e} I_{3\times3} + \left(q_{ve}\right)\times\right)\omega^0 \tag{18}$$

Substituting the first derivative of (5) and (17) into (16) yields

$$\dot{V} = \omega^{0T} I_s \left(I_s^{-1}\left(-\omega \times I_s \omega + \tau_c + c_d\right)\right) + \omega^{0T} I_s \dot{A}(q)\Omega^0 + K_p q_{ve}^T \omega^0 \tag{19}$$

where c_d is the environmental disturbances torque.

By substituting the controller (12) into (19), we can further obtain

$$\dot{V} = \omega^{0T} I_s \left(-\omega \times I_s \omega - K_p q_{ve} - K_d \omega^0 + \omega \times I_s \omega + c_d\right) + K_p q_{ve}^T \omega^0 = \omega^{0T}\left(K_d \omega^0 - c_d\right) \tag{20}$$

For $\dot{V} \leq 0$, the environmental disturbance c_d must satisfy the following inequality:

$$\left\|c_d\right\|_2 \leq \left\|K_d \omega^0\right\|_2 \tag{21}$$

3.2 Backstepping Controller

Backstepping control was developed by Kanellakopoulos [21]. This technique is a recursive method of synthesizing nonlinear control laws; it is based on Lyapunov stability.

Backstepping control's fundamental concept is to fragment the system into a set of subsystems of decreasing order that are stable in the Lyapunov sense, which offers them with asymptotic global stability robustness. In other words, it is a multistep technique. A virtual control is computed at each step to guarantee that the system will converge to its equilibrium state. This can be done by using Lyapunov functions, which guarantee the stabilization of each synthesis step.

Consider a system in the following form:

$$\dot{x}_1 = g_1 x_2 + f_1 \tag{22a}$$

$$\dot{x}_2 = g_2 x_3 + f_2 \tag{22b}$$

$$\vdots$$

$$\dot{x}_n = g_n U + f_n \tag{22c}$$

- Step 1
- We choose the first variable $z_1 = x_1$. Additionally, the second variable $z_2 = x_2 - \alpha_1$, with α_1 defined as a virtual control law.
- The expression for the time derivative of z_1 is

$$\dot{z}_1 = \dot{x}_1 = g_1 z_2 + g_1 \alpha_1 + f_1 \tag{23}$$

- To establish α_1, we construct the following Lyapunov function:

$$V_1(z_1) = \frac{1}{2} z_1^T z_1 \tag{24}$$

- The expression of the derivative of the Lyapunov function presented in Eq. (24) is

$$\dot{V}_1 = z_1^T \dot{z}_1 = z_1^T g_1 z_2 + z_1^T (g_1 \alpha_1 + f_1) \tag{25}$$

- A judicious choice of α_1 would make V_1 negative and ensure the stability of the subsystem described by Eq. (22a). Therefore, α_1 is chosen as follows:

$$\alpha_1 = \frac{1}{g_1}(-k_1 z_1 - f_1) \tag{26}$$

- We substitute α_1 in V_1, and in z_1, we find

$$\dot{V}_1 = -z_1^T k_1 z_1 + z_1^T g_1 z_2 \tag{27}$$

$$\dot{z}_1 = g_1 z_2 - k_1 z_1 \tag{28}$$

- For global stability, the last term $z_1^T g_1 z_2$ in $\dot{V}_1$ will be eliminated in the next step.
- Step 2
- In this step, we use a new variable $z_3 = x_3 - \alpha_2$:

$$\dot{z}_2 = \dot{x}_2 - \dot{\alpha}_1 = g_2 x_3 + f_2 - \dot{\alpha}_1 \tag{29}$$

- The Lyapunov function V_2 for the second subsystem is chosen as follows:

$$V_2(z_1, z_2) = V_1(z_1) + \frac{1}{2} z_2^T z_2 \tag{30}$$

- The expression of the derivative of the second Lyapunov function presented in Eq. (30) is

$$\dot{V}_2 = \dot{V}_1 + z_2^T \dot{z}_2 = \dot{V}_1 + z_2^T (g_2 x_3 + f_2 - \dot{\alpha}_1) \tag{31}$$

$$\dot{V}_2 = -z_1^T k_1 z_1 + z_2^T (g_1 z_1 + g_2 x_3 + f_2 - \dot{\alpha}_1) \tag{32}$$

$$\dot{V}_2 = -z_1^T k_1 z_1 + z_2^T g_2 z_3 + z_2^T (g_1 z_1 + g_2 \alpha_2 + f_2 - \dot{\alpha}_1) \tag{33}$$

- To make $\dot{V}_2$ negative, the virtual control law α_2 is chosen as follows:

$$\alpha_2 = \frac{1}{g_2}(-k_2 z_2 - g_1 z_1 - f_2 + \dot{\alpha}_1) \tag{34}$$

- We substitute α_2 in V_2, and in z_2, we find

$$\dot{V}_2 = -z_1^T k_1 z_1 - z_2^T k_2 z_2 + z_2^T g_2 z_3 \tag{35}$$

$$\dot{z}_2 = g_2 z_3 - k_2 z_2 - g_1 z_1 \tag{36}$$

- For global stability, the last term $z_2^T g_2 z_3$ in $\dot{V}_2$ will be eliminated in the next step.
- nth Step
- This is the last design step, since the final command U appears in the derivative of z_n:

$$\dot{z}_n = \dot{x}_n - \dot{\alpha}_{n-1} = g_n U + f_n - \dot{\alpha}_{n-1} \tag{37}$$

- The global Lyapunov function is expressed by

$$V_n\left(z_1,z_2\ldots z_n\right)=V_{n-1}\left(z_1,z_2\ldots z_{n-1}\right)+\frac{1}{2}z_n^T z_n \tag{38}$$

- The derivative of V_n is

$$\dot{V}_n=\dot{V}_{n-1}+z_n^T\dot{z}_n \tag{39}$$

- which leads to

$$\dot{V}_n=\sum_{i=1}^{n-1}\left(-z_i^T k_i z_i\right)+z_n^T\left(g_{n-1}z_{n-1}+g_n U+f_n-\dot{\alpha}_{n-1}\right) \tag{40}$$

- To make the derivative of the global Lyapunov function negative, the control law U is chosen as follows:

$$U=\frac{1}{g_n}\left(-k_n z_n-g_{n-1}z_{n-1}-f_n+\dot{\alpha}_{n-1}\right) \tag{41}$$

- Substituting U into V_n and into z_n, we find

$$\dot{V}_n=\sum_{i=1}^{n}\left(-z_i^T k_i z_i\right) \tag{42}$$

$$\dot{z}_n=-k_n z_n-g_{n-1}z_{n-1} \tag{43}$$

- Now, we present the application of the backstepping command to the satellite model. The two steps of the backstepping algorithm are as follows [21]:
- *Step 1*
- The first and second variables of backstepping are defined as

$$\mathbf{z_1}=\mathbf{x_1}=\mathbf{q_e}=\mathbf{q_c q} \tag{44}$$

$$\mathbf{z_2}=\mathbf{x_2}-\alpha_1 \tag{45}$$

- where
- $\mathbf{q_c}=[q_{c1}\quad q_{c2}\quad q_{c3}\quad q_{c4}]$ is the control quaternion.
- $\mathbf{x_2}=\omega_s^o$.

- $\boldsymbol{\alpha}_1$ is a virtual control law.
- The derivative of $\mathbf{z}_1$ is

$$\dot{\mathbf{z}}_1 = \dot{\mathbf{x}}_1 = \dot{\mathbf{q}}_e = \mathbf{q}_c\left(\frac{1}{2}\Lambda(q)\omega_s^o\right) \tag{46}$$

- The initial Lyapunov function is expressed as

$$V_1(\mathbf{z}_1) = z_1^T\mathbf{z}_1 \tag{47}$$

- The time derivative of V_1 is

$$\dot{V}_1(\mathbf{z}_1) = 2z_1^T\dot{z}_1 = z_1^T\boldsymbol{G}(q)\mathbf{z}_2 + z_1^T\boldsymbol{G}(q)\alpha_1 \tag{48}$$

- where

$$\boldsymbol{G}(q) = \mathbf{q}_c\Lambda(q) \tag{49}$$

- For generating $\dot{V}_1$ negative, $\boldsymbol{\alpha}_1$ is chosen as

$$\alpha_1 = -\mathbf{k}_1\boldsymbol{G}(q)^T\mathbf{z}_1 \tag{50}$$

- where
- $\mathbf{k}_1$ is a positive gain matrix.
- The time derivative of V_1 becomes

$$\dot{V}_1(\mathbf{z}_1) = -z_1^T\boldsymbol{G}(q)\mathbf{k}_1\boldsymbol{G}(q)^T\mathbf{z}_1 + z_1^T\boldsymbol{G}(q)\mathbf{z}_2 \tag{51}$$

- The term $z_1^T\boldsymbol{G}(q)\mathbf{z}_2$ will be eliminated in the next step.
- *Step 2*
- The time derivative of $\mathbf{z}_2$ is expressed as

$$\dot{\mathbf{z}}_2 = \dot{\omega}_s^o - \dot{\alpha}_1 \tag{52}$$

- where

$$\omega_s^o = \omega_s^I - \boldsymbol{A}\omega_o \tag{53}$$

- When we substitute Eq. (53) into Eq. (52), we get

$$\dot{\mathbf{z}}_2 = \dot{\omega}_s^I - \dot{A}\omega_o - \dot{\alpha}_1 \tag{54}$$

$$\mathbf{I}\dot{\mathbf{z}}_2 = \mathbf{I}\dot{\omega}_s^I - I\dot{A}\omega_o - I\dot{\alpha}_1 \tag{55}$$

- The second Lyapunov function is chosen as

$$V_2(\mathbf{z}_1, \mathbf{z}_2) = V_1(\mathbf{z}_1) + \tfrac{1}{2} z_2^T I z_2 \tag{56}$$

- Its time derivative is

$$\dot{V}_2(\mathbf{z}_1, \mathbf{z}_2) = \dot{V}_1(\mathbf{z}_1) + z_2^T I \dot{\mathbf{z}}_2 \tag{57}$$

$$\begin{aligned} \dot{V}_2(\mathbf{z}_1, \mathbf{z}_2) &= -z_1^T G(q)\mathbf{k}_1 G(q)^T \mathbf{z}_1 + z_1^T G(q)\mathbf{z}_2 \\ &+ z_2^T \left[C_{ext} + I\omega_s^I \times \left(I\omega_s^I + h \right) - \dot{h} \right] - I\dot{A}\omega_o - I\dot{\alpha}_1 \end{aligned} \tag{58}$$

- To make $\dot{V}_2$ negative, the control law $\dot{h}$ is chosen as

$$u = \dot{h} = \mathbf{k}_2 \mathbf{z}_2 + G(q)^T \mathbf{z}_1 + \left(C_{ext} + I\omega_s^I \times \left(I\omega_s^I + h \right) \right) - I\dot{A}\omega_o - I\dot{\alpha}_1 \tag{59}$$

4 Results of the Simulation

In this section, numerical simulations are presented for the CubeSat's attitude stabilization problem to confirm the effectiveness of the utilized controllers. The simulations were set up to run for 300 s with a 0.01-s sampling rate. The suggested SPD and backstepping controllers are directly compared in terms of performance (Fig. 1).

The results of the attitude control are validated using 1HOPSAT 12U CubeSat parameters, which have the following inertia matrix:

$$\mathbf{J} = \begin{bmatrix} 0.2853 & 0 & 0 \\ 0 & 0.2853 & 0 \\ 0 & 0 & 0.1467 \end{bmatrix} \left[\text{kg}\,\text{m}^2 \right]$$

The satellite orbital speed is $\boldsymbol{\Omega}_0 = [0; -0.0011; 0]$ rad/s.

- The initial attitude is selected as $\boldsymbol{q}(0) = [0.9981; 0.0435; -0.0435; 0.0019]$.
- The initial angular velocity as $\omega(0) = [0.005; 0.005; -0.005]$ rad/s.

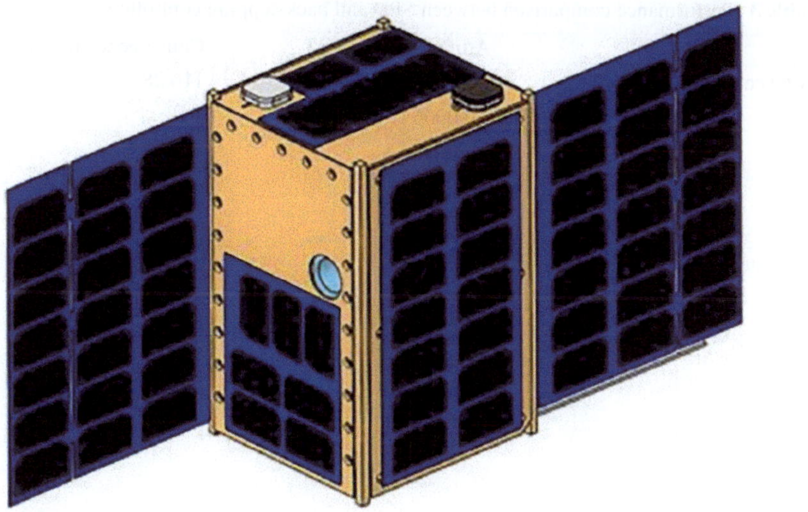

Fig. 1 1HOPSAT (first-generation high optical performance satellite)

Table 1 SPD Parameters

k_P	0.08
k_d	0.4

Table 2 Backstepping parameters

k_1	0.01 eye(3)
k_2	8 eye(3)

- The desired attitude is chosen as $\boldsymbol{E}_d = [15; -15; -24]$ deg.
- The gravitational constant of the Earth $\mu = 3.9859\ 10^{14}\ \mathrm{m}^3\ \mathrm{s}^{-2}$
- The mean orbit radius $r_0 = 6757$ km.

Tables 1 and 2 give the SPD and backstepping controller parameters, respectively.

The simulation results of the similar proportional-derivative controller are presented below:

And the simulation results of the backstepping controller are presented as follows:

Figure 10 presents the roll, pitch, and yaw angles for SPD and BS controllers.

A more detailed analysis can be found using the computation of the root mean square error (RMSE) values and convergence time. The results are presented in Table 3.

Figures 2, 3, 4, 5, 6, 7, 8, 9, and 10 present the simulation results for the backstepping control and the similar proportional-derivative control. It is noted that both control laws yield satisfactory results, effectively ensuring that the attitude converges to the desired state.

Table 3 Performance comparison between SPD and backstepping controllers

	Attitude RMS (deg)	Convergence time (s)
SPD controller	$\begin{bmatrix} 0.0006 \\ 0.0017 \\ 0.0008 \\ 0.0019 \end{bmatrix}$	116.28
Backstepping controller	$\begin{bmatrix} 0.0015 \\ 0.0039 \\ 0.0005 \\ 0.0042 \end{bmatrix}$	75.92

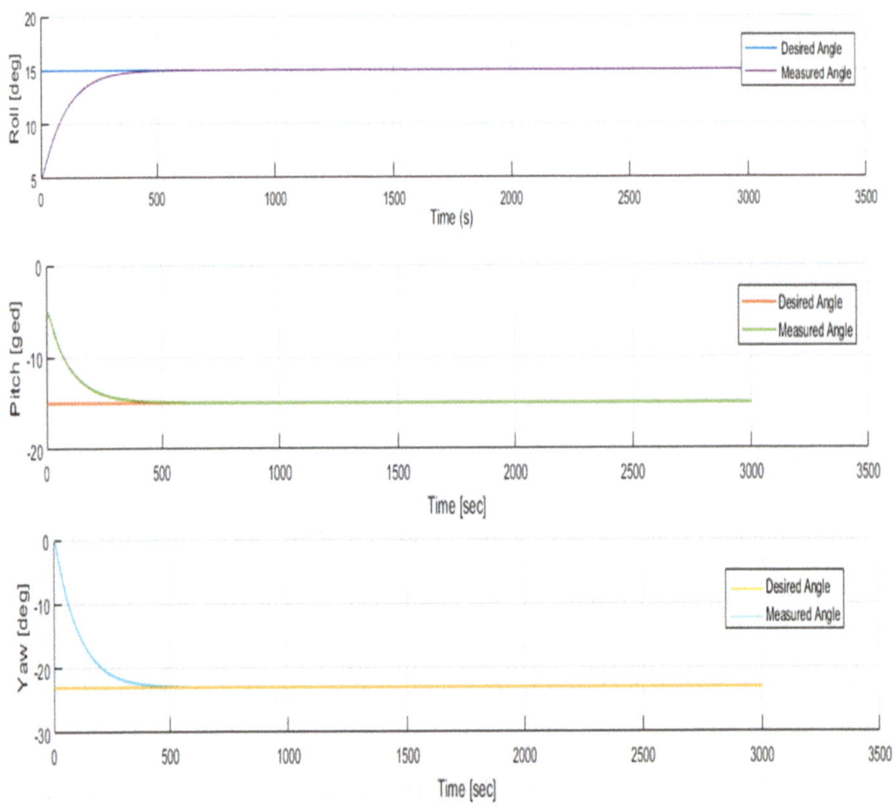

Fig. 2 Euler angles

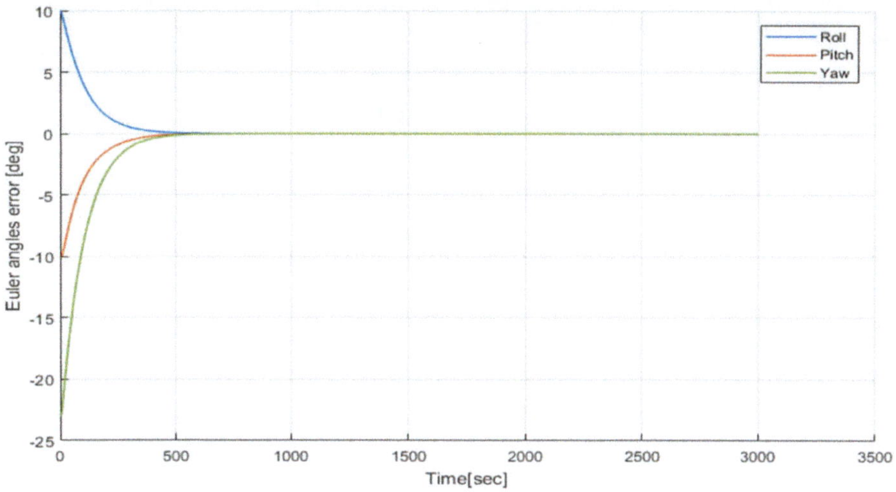

Fig. 3 Euler angle errors

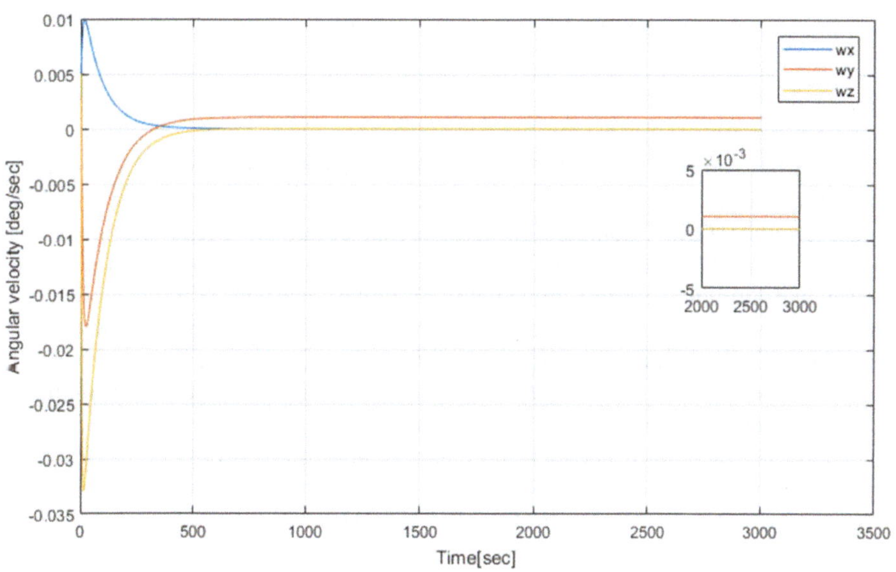

Fig. 4 Angular velocity

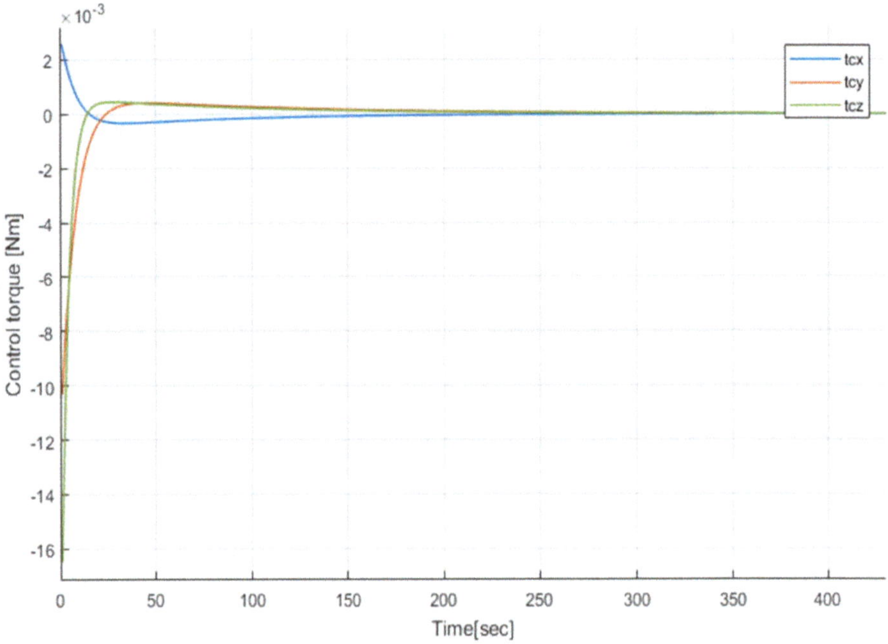

Fig. 5 Control torque

It can be seen that within 116 s, the SPD controller smoothly and successfully converges to the desired attitude angles. It retains a very low error for the Euler angle (around 0.0019°) compared with the second controller. The proportional component of the SPD controller swiftly addresses the existing error between the desired and actual attitude, offering immediate corrective measures. Simultaneously, the derivative component anticipates future error changes, mitigating oscillations or overshoots by assessing the error's rate of change. Moreover, the BS controller takes the shortest time for its attitude convergence compared to the SPD controller, with a convergence time of 75.92 s and an acceptable attitude error.

Therefore, the choice between SPD and BS algorithms should be based on the specific requirements of the application. If precise control of the attitude is more important, SPD would be the preferred choice. However, if faster convergence is required, BS would be more suitable.

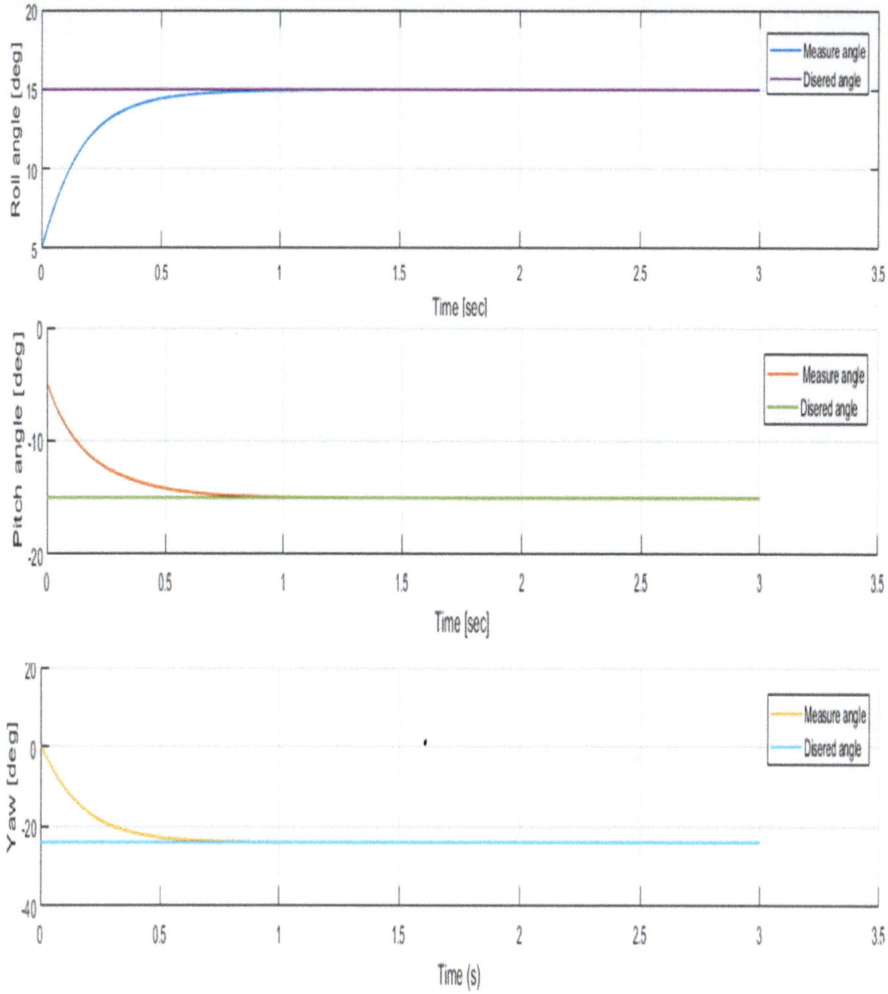

Fig. 6 Euler angles

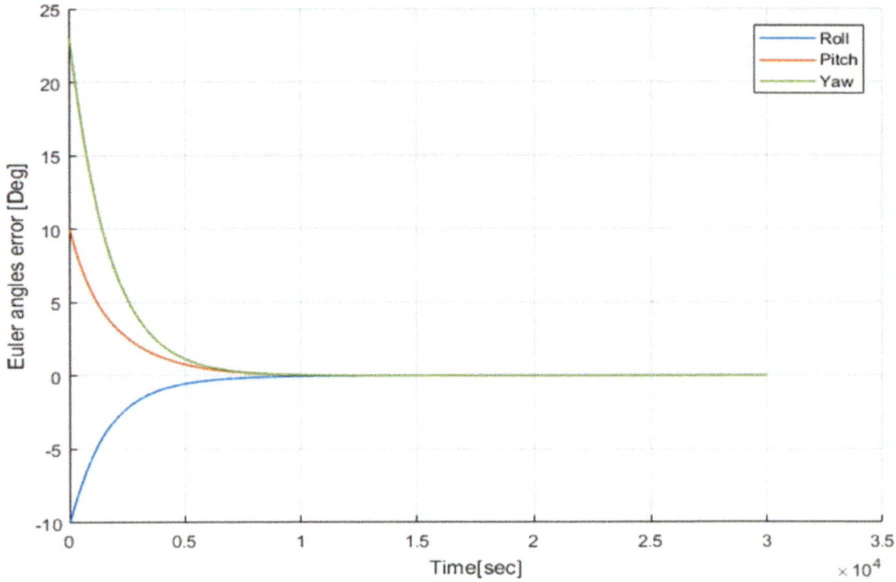

Fig. 7 Euler angle errors

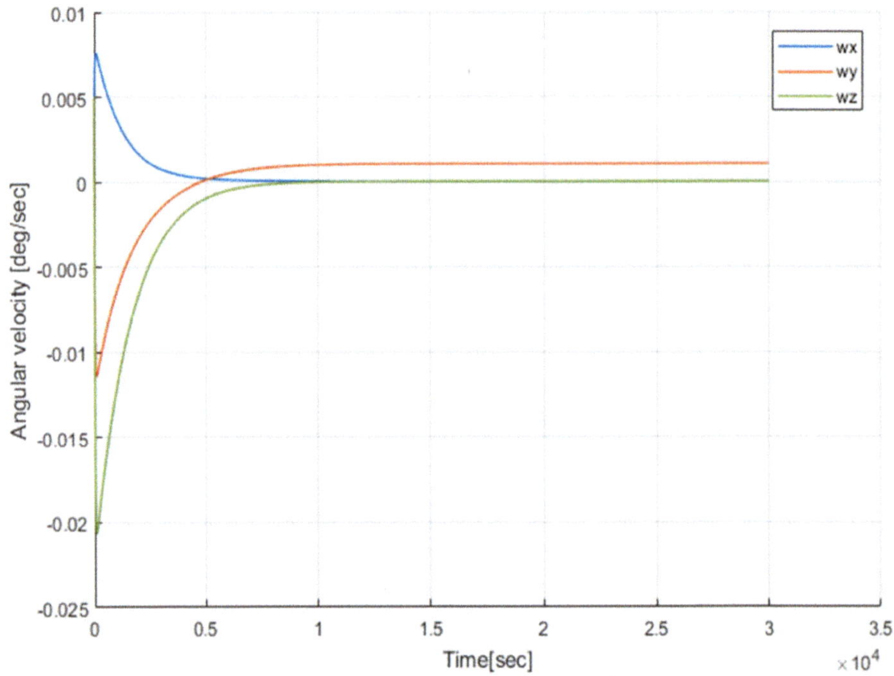

Fig. 8 Angular velocity

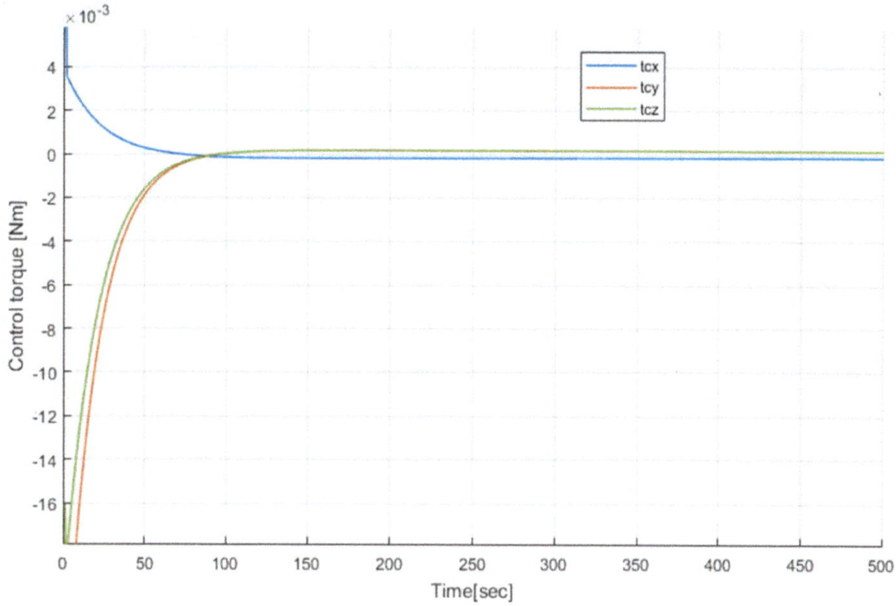

Fig. 9 Control torque

5 Conclusion

In this study, we used simulations to show the efficacy of the SPD and backstepping controllers in CubeSat attitude control. The backstepping controller functioned exceptionally well at controlling nonlinearities and disturbances. The SPD method succeeded exceptionally well in precision with reduced attitude root mean square (RMS) values, according to the precision vs. speed trade-off analysis, which makes it appropriate for applications requiring accuracy. However, in situations where rapid response is more important than accurate attitude control, the backstepping algorithm's faster convergence times are favorable. Therefore, the choice between SPD and backstepping algorithms should be based on the specific requirements of the application: if accurately controlling the attitude is more important, SPD would be the preferred choice, but if rapid convergence is required, backstepping would be more appropriate.

AI Use Disclosure Statement The authors did not use any generative AI or AI-assisted technologies in the creation of this manuscript. The authors used generative AI technologies only for proofreading. This includes minor corrections to grammar, spelling, and syntax. No AI was used for ideation, data analysis, or writing original content.

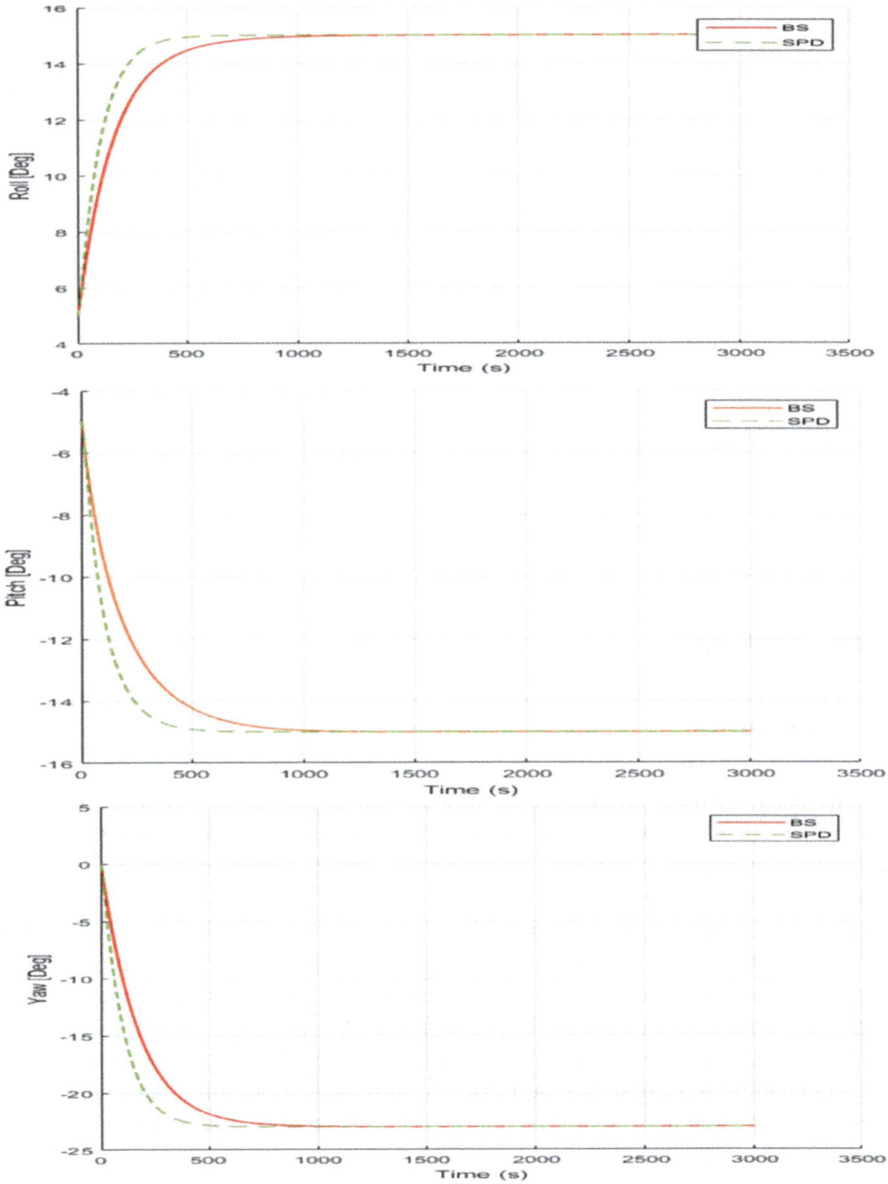

Fig. 10 Euler angles

References

1. Boussadia H, Si Mohammed MA, Boughanmi N, Meche A, Bellar A (2022) A combined configuration (αβ filter-TRIAD algorithm) for spacecraft attitude estimation based on in-Orbit Flight Data. Aerosp Syst 5:223–232
2. Boussadia H, Si Mohammed MA, Meche A, Boughanmi N, Slimane A, Bellar A (2023) Estimation of satellite attitude dynamics and external torques via mixed Kalman/H-infinity filter under inertia uncertainties. Aerosp Syst 6:633–640
3. Boussadia H, Mohammed AS, Boughanmi N, Bellar A (2017) Adaptive backstepping control for microsatellite under inertia uncertainties. In: Proceedings of the 39th IEEE conference on recent advances in space technologies, Istanbul, Turkey, 19–22 June 2017
4. Mohammed MAS, Boussadia H, Bellar A, Adnane A (2016) Adaptive backstepping control for three axis microsatellite attitude pointing under actuator faults. In: Proceedings of the 13th European workshop on advanced control and diagnosis (ACD 2016), Lille, France, 17–18 November 2016
5. Cortiella A, Vidal D, Jané J, Juan E, Olivé R, Amézaga A, Munoz JF, Carreno-Luengo PVH, Camps A (2016) 3cat-2: attitude determination and control system for a GNSS-R earth observation 6U CubeSat mission. Eur J Remote Sens 49(1):759–776
6. Cipro F (2020) CubeSat attitude and position control systems based on LPV and Sliding Mode methods. PhD thesis, Politecnico di Torino
7. Murugan P, Agrawal Y (2020) Small satellites applications, classification and technologies. Int J Sci Res 9(7):1682–1687
8. Fortescue P, Swinerd G, Stark J (2011) Spacecraft systems engineering. Wiley
9. Pan S, Zhu D, Liu S, Xu K, Dai Y, Wang T, Liu J, Zhu N, Xue Y, Liu N (2015) Satellite payloads pay off. IEEE Microw Mag 16(8):61–73
10. Nguyen HH, Nguyen PS (2020) Communication subsystems for satellite design. In: Satellite systems-design, modeling, simulation and analysis. IntechOpen
11. Ridolfi G, Mooij E, Corpino S (2009) A system engineering tool for the design of satellite subsystems. In: AIAA modeling and simulation technologies conference, p 6037
12. Efendioğlu G (2019) Design of Kalman filter based attitude determination and control algorithms for a Leo satellite. Master's thesis, Middle East Technical University
13. Santiago MC et al (2018) Cubesat attitude control system based on embedded magnetorquers in photovoltaic panels
14. Sharma T (2021) Modelling and simulation of attitude and orbital control for foresail-2
15. Kaplan C (2006) Leo satellites: attitude determination and control components; some linear attitude control techniques. Master's thesis, Middle East Technical University
16. Mohanasundaram S (2017) Návrh a simulace řídicích zákon pro stabilizaci satelit. Master's thesis, České vysoké učení technické v Praze, Vypočetní a informační centrum
17. Losa D (2007) High vs low thrust station keeping maneuver planning for geostationary satellites. PhD thesis, École Nationale Supérieure des Mines de Paris
18. Ravandoor K (2008) Design and simulation of software based attitude determination system for the CubeSat UWE-2
19. Wertz JR (1986) Spacecraft attitude determination and control. D. Reidel Publishing Company, Boston, MA
20. Sidi M (1997) Spacecraft dynamics and control. Cambridge University Press
21. Kanellakopoulos I, Kokotovic PV, Stephen Morse A (1991) Systematic design of adaptive controllers for feedback linearizable systems. In: 1991 American control conference. IEEE, pp 649–654

Next-Generation Simulation Ecosystems Bridging Digital Twin and Omniverse Technologies

Amir Faraji, Shima Homayoon Arya, Zahra Alirezaei, Hannan Abagheri Mahabadi, and Bahar Mehdizadeh

Abstract Engineering and construction projects often deal with many complicated issues and keep running into problems like disconnected work processes, difficulty in sharing information between different systems, and weak teamwork across various fields. Digital twin (DT) technology has become a big change in how these challenges are handled. It connects the real world with the digital world by allowing real-time tracking, smart predictions, and managing projects throughout their entire life. Even though DT started in the aerospace industry, it has quickly become popular in construction because of the need for more sustainable and people-focused building practices. In parallel, NVIDIA Omniverse has introduced an open platform based on Universal Scene Description (USD) for immersive, real-time collaboration and simulation. This chapter explores how DTs are classified across complexity, life cycle phases, and hierarchical perspectives and evaluates their applications in pre-construction, construction, and post-construction stages. It also looks at Omniverse as a system that can grow and support many users at the same time. This system brings in AI-powered simulations, real-time graphics, and teamwork features. The study shows that integrating advanced twins with simulation, visualization tools, and collaboration features can transform the construction sector into a more con-

---

A. Faraji
Construction Project Management Department, Faculty of Architecture, Khatam University, Tehran, Iran

School of Engineering, Design and Built Environment, Western Sydney University, Penrith, NSW, Australia
e-mail: a.faraji@khatam.ac.ir; a.faraji@westernsydney.edu.au

S. H. Arya · Z. Alirezaei · H. A. Mahabadi
Construction Project Management Department, Faculty of Architecture, Khatam University, Tehran, Iran
e-mail: sh.homayoon@khatam.ac.ir; z.alirezaei.9451@khatam.ac.ir; H.Abagheri.8754@khatam.ac.ir

B. Mehdizadeh (✉)
School of Engineering, Design and Built Environment, Western Sydney University, Penrith, NSW, Australia
e-mail: B.Miyandehi@westernsydney.edu.au

M. Mirrashid et al. (eds.), *Optimization and Adaptive Control Algorithms*, Emerging Trends in Mechatronics, https://doi.org/10.1007/978-981-95-8387-4_4

nected and predictive system. This approach improves efficiency, sustainability, and innovation, aligning with the principles of Industry 5.0, which emphasize human-machine collaboration. The results also reveal that combining DTs with Omniverse reduces inefficiencies and creates new opportunities for digital collaboration, immersive visualization, and smart infrastructure management.

Keywords Digital twin (DT) · Omniverse · Simulation · Construction industry Construction 5.0

1 Introduction

Traditional workflows have usually relied on isolated systems, with different software applications operating independently and using their own file formats and data structures. This lack of compatibility makes collaboration more difficult, often requiring repeated file conversions and manual data synchronization, which can slow down processes and reduce overall efficiency. Recently, the transition to remote and distributed work has resulted in more teams working together concurrently on shared projects, underscoring the growing demand for tools and technologies that facilitate real-time collaboration, irrespective of geographical location. Real-time collaboration is especially vital in the realm of 3D content creation, where numerous artists, designers, and developers must collaborate effortlessly to construct intricate, dynamic, and highly detailed virtual environments. These collaborative tools not only optimize the workflow but also boost creativity by enabling immediate feedback and synchronized efforts among teams [1, 2].

Digital twins (DTs) are a twenty-first-century concept that has attracted rapidly growing interest. Originally developed for the automotive and aerospace sectors, DTs are now being applied to almost any physical product or artifact, whether tangible or intangible. DTs are virtual replicas of real or planned physical objects, offering detailed representation through data rather than existing fully or not at all [3, 4]. The level of detail and how often a digital twin (DT) is updated with its physical counterpart depend on the purpose and value it provides. While DTs themselves have gained a lot of attention, understanding the environment they operate in is just as important. By including environmental data, different DTs can work together in shared digital spaces, forming what is called a DT Metaverse, where collaboration and complex interactions are possible. On top of that, simulation is a key part of this evolution, allowing predictions and testing in these virtual spaces. Future scenarios can be anticipated through simulation, even if only probabilistically [5, 6]. While possibilities have always been imagined and modeled by humans, what is new is that these simulations can now be performed externally using DTs, and predictive power is thus extended beyond the human mind [7].

The idea of the Metaverse, a term created from combining "meta" meaning virtual or beyond and "verse" from universe, has drawn increasing attention in recent years. Much of the growing attention to the Metaverse comes from recent technological breakthroughs, including 5G, blockchain (BCT), augmented reality (AR) and virtual reality (VR), and wearable headsets [8, 9]. The idea itself was first brought into the spotlight by Neal Stephenson's novel Snow Crash, where it was imagined as a three-dimensional digital world in which people connect and interact through avatars, transforming the Internet into a fully immersive space. Such a development holds the possibility of reshaping human communication as well as our relationship with digital systems [10–12]. Pixar introduced the Universal Scene Descriptor (USD) in 2016 as an open-source standard for creating and sharing 3D scenes. Building on this, NVIDIA launched the Omniverse platform in 2020, designed to work directly with USD. Omniverse supports large-scale scene creation, real-time ray tracing, and collaborative development. Combined, USD and Omniverse provide a strong framework for seamless collaboration, visualization, and data integration across different platforms [2].

2 Digital Twin

The gap between the digital and physical realms hampers efficiency and teamwork in the construction sector. DT technology, which creates a virtual model of a physical asset for purposes such as simulation, monitoring, analysis, and various life cycle tasks, presents a viable solution to address these issues [13]. DTs serve as virtual replicas of tangible assets, merging digital representations with communication capabilities, which allows them to function as smart elements within the Internet of Things (IoT) and related services [14]. Additionally, they can be perceived as digital models that mirror real-world entities, emphasizing the characteristics and actions of their physical equivalents [15]. A DT signifies a virtual and authentic depiction of tangible assets, processes, or systems situated in either the constructed or the natural surroundings [16]. A two-way exchange of cohesive data is frequently seen as a fundamental characteristic of a DT. However, it is not necessary to include every digital element generated throughout the creation of a physical counterpart [17]. Rather, a DT integrates solely the critical information needed to achieve its specific goals [18].

The phrase DT was first introduced to the public through NASA's integrated technology road map. The origins of this concept can be linked to NASA's Apollo program, which pioneered the notion of "twins" by constructing two identical spacecraft, one launched for the mission and the other retained on Earth to simulate and observe the conditions encountered in space. In 2003, Michael Grieves from the University of Michigan articulated the notion of a "digital counterpart of a physical product," which is widely acknowledged as the initial formal presentation of the DT idea. Subsequently, in 2006, he introduced the term "product avatar" to create an information management framework that facilitates product-centric, two-way data exchanges. Grieves further expanded on this concept in a 2014 white paper, reinforcing the foundational principles of DT theory [15]. In recent times, the study of DTs has grown significantly, evidenced by an increase in associated publications.

This growth underscores the unavoidable shift toward a more profound connection and amalgamation of the physical and digital realms [19].

DTs originated in the manufacturing industry as a method to monitor and oversee products throughout their complete life cycle. Currently, they are extensively utilized in technologically advanced fields such as manufacturing, automotive, and aerospace sectors. DTs provide a comprehensive insight into the functioning of infrastructure and its interactions with the surrounding environment, aiding governments and industries in making better-informed choices for future advancements. A technology trend analysis identified DT as one of the leading 4 emerging technologies among a total of 15 [19]. While the initial expenses of adopting DTs can be considerable, they provide considerable long-term financial benefits [8]. Advantages encompass enhanced quality, lower operational expenses, and heightened revenue. In summary, DTs boost efficiency by conserving time, energy, and resources [16].

As shown in Fig. 1, the primary features of DTs include real-time mirroring, which ensures a precise and synchronized connection between the physical and digital worlds; interaction and convergence, allowing DTs to combine historical and real-time data to create cohesive simulations and enable seamless two-way interactions between physical and virtual realms; and self-evolution, which provides the ability to continuously update their virtual representations for ongoing improvement and adaptation [19]. Figure 1 provides a comprehensive overview of the DT concept, illustrating how this technology works starting from collecting, processing, generating, and managing. Moreover, the other part of this figure illustrates the timeline of this technology, starting from 1970. Additionally, supporting technologies help DT to improve abilities in various cases. It shows how DT integrates with supporting technologies such as virtual modeling, data processing, simulation, and smart devices. On the other hand, this figure also highlights DT's role among different project phases such as preconstruction, construction, and post-construction. Some key features such as safety management, cost control, quality monitoring, and sustainability are also depicted.

2.1 Overview of DT Types

In recent times, the embrace of and focus on DTs have surged notably, driven by their capacity to transform product design, manufacturing, and operational processes. Since 2016, there has been a substantial increase in publications regarding DTs, reflecting heightened research enthusiasm and industrial emphasis [20]. DTs have been recognized as a pivotal technology by Gartner [21], designated as an emerging technology for the aerospace and defense sectors by Lockheed Martin [22], and emphasized by Forbes [23] as one of the key technologies shaping the upcoming decade, highlighting their increasing importance for innovation in various industries. Juniper Research estimates that the worldwide DT market was expected to hit USD 12.7 billion by 2021, and this swift growth has led companies in all fields to invest more heavily in this technology. To successfully utilize DT technology across various industries, it is vital to grasp the advantages it can

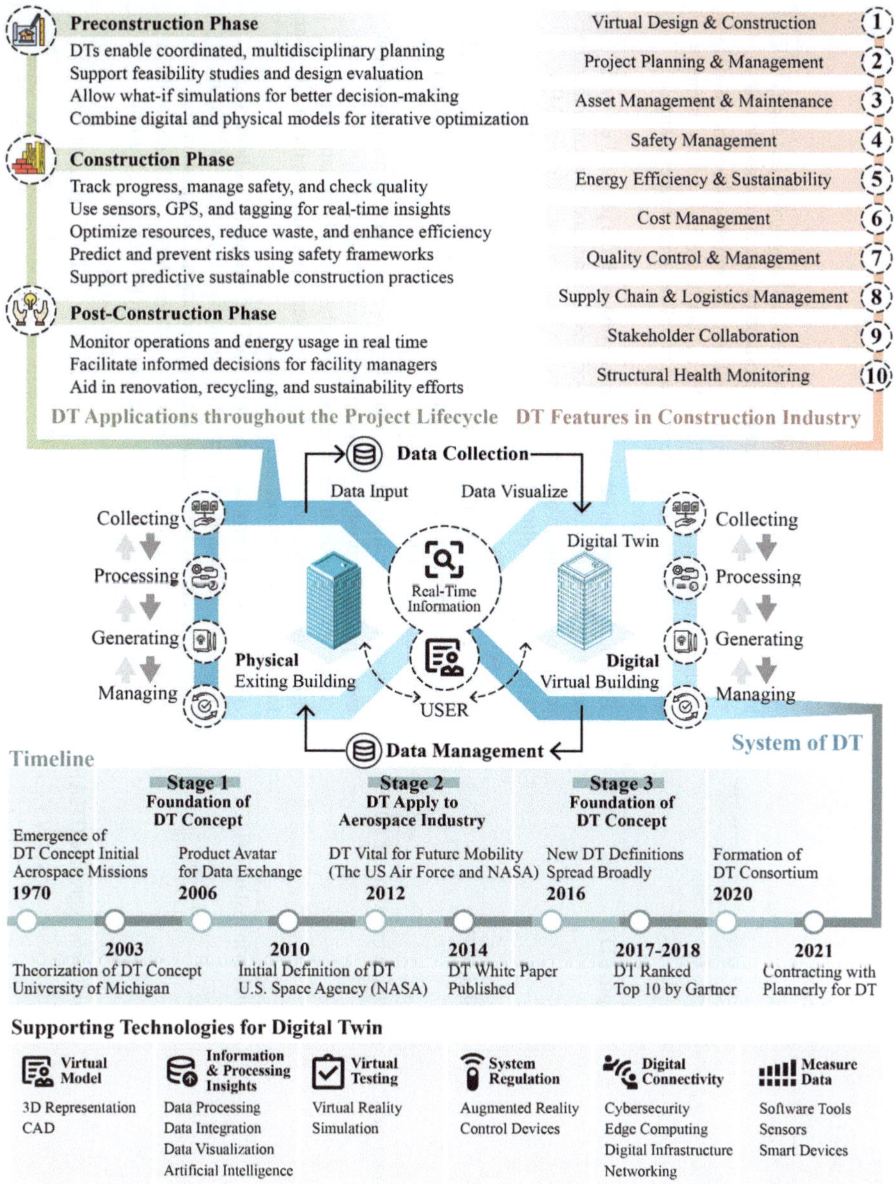

Fig. 1 A comprehensive overview of DT

provide to the organization. Furthermore, identifying the unique features and categories of DTs is important for choosing the version that most closely matches business goals and enhances potential gains. Consequently, a thorough comprehension of the various types of DTs is essential for selecting the most suitable type and maximizing their strategic and financial effects [20]. Table 1 presents the

Table 1 Classification of DTs

No.	Categorization type	Reference	Category title	Definition
1	Intricacy of the virtual model [24]	Pre-DT	At this point, the DT is created before the actual physical object and is utilized to aid in decision-making, acting as a prototype to reduce possible risks
		DT	This phase signifies the second tier, wherein information from the tangible product is integrated. It is utilized throughout the design and development decision-making stages of the product life cycle, with data exchange happening in both directions
		Adaptive DT	The DT establishes an interactive connection with the physical entity. It possesses the ability to learn in a prioritized manner and can remember the preferences of human operators via supervised machine learning (ML). Furthermore, it facilitates immediate decision-making and planning in operational contexts
		Intelligent DT	Unlike the adaptive DT, this DT integrates unsupervised ML features, which boosts its independence. Additionally, it provides improved precision and more effective system evaluation
2	Time of development [25, 26]	Digital twin prototype (DTP)	Developed in the design stage, the DTP acts as a virtual representation that encompasses the product's physical attributes, expected performance, and production methods. It is subjected to simulations and evaluations to determine the quality of the final physical version while also integrating plans for ongoing maintenance and assistance
		Digital twin instance (DTI)	After the physical product is created, the DTI acts as its "as-built" digital equivalent, maintaining a connection throughout the product's lifespan. It records deviations from manufacturing, operational information, and performance records. The bidirectional data exchanges facilitate monitoring, forecasting, and digital experimentation, thereby increasing trust in simulations and lessening reliance on physical testing
		Digital twin aggregate (DTA)	At an elevated level, the DTA brings together all DTIs, merging extensive data for the purposes of analysis, forecasting, and AI/ML applications. The combination of vast datasets improves predictive precision, though the outcomes are still vulnerable to possible model bias

3	Hierarchical perspective [27]	Unit level	Denotes the most fundamental entity engaged in production, concentrating on the functional, geometric, behavioral, and operational models of its tangible equivalent
		System level	Signifies a consolidation of unit-level DTs, with each unit-level DT aligning with a part of the complete system
		System of systems (SoS) level	Illustrates the linkage of several system-level DTs to create a system of systems, incorporating diverse stages throughout the product life cycle
4	Use and decision horizon [28]	High-performance real time (HPRT)	Utilized in intricate, data-heavy industries like aerospace, manufacturing, and self-driving vehicles, concentrating on real-time information, simulation, and artificial intelligence (AI)
		Long-term decision support (LTDS)	Mainly applied in the constructed environment, highlighting data representation, validation, visualization, and decisions driven by policy considerations
5	Built environment (based on LTDS) [28]	Building digital twins (BDTs)	BDTs to assist in overseeing facilities and managing their life cycle
		Architectural digital twins (ADTs)	ADTs to facilitate design, strategizing, and cooperation among involved parties
		Urban/city digital twins (UDTs/CDTs)	UDTs/CDTs to enhance urban development, management, transit, and community services via data integration powered by IoT technology

classification of DT technology, indicating the criteria used for categorization and outlining the key elements included in each category.

At present, the growing importance of DTs, which serve to significantly enhance the capabilities and functionalities of information models, is primarily driven by advancements in various digital technologies. These include innovations in architecture, ongoing efforts toward standardization, emergence of new use cases, improvement of interactions among users [29], and development of inventive business models that are made possible through the implementation of DT technology [27].

2.2 The Role of DT in the Construction Industry

In the construction sector, a lack of adequate connection between the digital and physical domains, especially regarding data analysis and subsequent actions, has led to disjointed information, along with data redundancy and stagnation throughout various phases of the construction life cycle [13]. Significant obstacles faced by the industry encompass low efficiency, a negative public perception, restricted predictability, structural disunity, and inadequate research, development, and investment in innovation [15]. It diminishes management effectiveness and obstructs cooperation among employees, diverse professions, and various stages of construction [30]. Construction sites are notably fluid, experiencing constant alterations in equipment, materials, workforce, site organization, and other elements throughout different phases, which amplifies the necessity for coordination and collaboration. Consequently, the intricacy of overseeing and managing construction sites is greatly elevated [13].

For technological innovations to be practical and broadly embraced in the construction sector, they need to tackle the fundamental issues that pose significant challenges [31]. Investing in DTs can improve productivity via predictive analytics, simultaneously alleviating numerous ongoing problems within the industry [15]. A digital twin consists of a collection of virtual models that replicate a physical asset, offering insights and understanding comparable to those obtained from the actual object [18]. The fundamental components of DTs involve the connection of data that bridges the physical and digital realms, along with the two-way, dynamic engagement between tangible items and their corresponding virtual representations. A robust connection is established between the physical and digital elements, driven by ongoing and dynamic data interchange. This allows for real-time observation, modification, simulation, analysis, management, forecasting, and enhancement; it additionally facilitates validation and smooth collaboration during the entire project life cycle, effectively averting information silos and data redundancy. This method enables thorough oversight and administration of construction site conditions throughout the whole process [13].

Information is usually gathered through on-site sensors that consistently track alterations in the building and its surroundings [32], thereby updating the BIM model as needed [16]. Kritzinger introduced a categorization of DTs according to their degree of connectivity with the physical model. A digital model lacks any link to its physical equivalent; a digital shadow exhibits a one-way automated data transfer from the physical model to the digital model, while a DT facilitates a two-way automated data exchange, allowing the digital model to manage the physical asset. A practical illustration of the DT concept is its utilization on construction sites, where it can proficiently represent upcoming tasks and gather data specific to the site. DTs can be customized for particular goals during a project's life cycle, incorporating pertinent information to faithfully replicate the physical asset within a specified use case. While DTs have seen extensive use in the manufacturing sector, their influence in the construction industry has been relatively minimal [15]. Findings from the DT segment were categorized into themes aligned with the life cycle stages of a constructed asset (preconstruction, construction, and post-construction) to evaluate the possible benefits that DTs may offer at every phase [33]. This technology is capable of encompassing the complete life cycle of a building, which includes its design, construction, operation and maintenance, renovation, and eventual demolition.

2.2.1 DT Applications in Preconstruction Phase

During the planning and design phase, different disciplines and consultants usually carry out their tasks separately. This lack of integration can create serious challenges during the construction stage. Employing a DT in the planning process makes it possible to establish a coordinated, multidisciplinary approach that links people with urban design practices. Moreover, DT can help stakeholders evaluate the feasibility of proposed designs, identify the most suitable design alternative for the project, and investigate various scenarios through simulation to perform what-if analysis. DTs integrate the information model with the physical model of a product, allowing both models to be refined iteratively. This continuous optimization shortens the overall design process and lowers the risk of additional expenses caused by rework. A construction project typically progresses through a series of phases [34, 35]. Within the design and engineering stage, activities include inception, briefing, design, and detailed engineering [36].

Studies on the incorporation of DTs into preconstruction processes have concentrated on planning and design choices, evaluation of tenders, assessment of sustainability, and monitoring of environmental factors [33]. In the phase of architectural design, DTs have the capability to model the energy usage of different design alternatives, allowing for the identification of the most energy-efficient and eco-friendly option. DTs can enhance data retrieval, improve design efficiency, and aid in the effective allocation of resources throughout the preconstruction phase. This can aid in making decisions related to aspects like choosing materials, managing energy, purchasing, and assessing suppliers [15]. In the current age of eco-friendly construction and sustainable development, DTs hold the promise of enhancing sustainability goals throughout preconstruction processes [33]. Certain scholars have investigated the incorporation of particular digital technologies alongside DTs, whereas numerous studies have demonstrated that DTs can greatly enhance sustainable development [37].

2.2.2 DT Applications in Construction Phase

Throughout the construction stage of a building's life cycle, DTs have been utilized for tracking progress, managing safety, preparing for emergencies, overseeing overall construction operations, and handling waste. Additionally, they aid in quality assessment by evaluating geometric precision, confirming completeness, and directing essential modifications to building components. At present, various technologies are utilized in the construction industry [38], such as GPS for monitoring project advancement and tagging systems for tracking materials and personnel locations, along with intelligent sensors and sensor networks to evaluate construction quality [15]. The creation of DT platforms for construction safety has become crucial for prompt and informed decision-making, facilitating the derivation of practical safety insights from unrefined, value-producing data.

In a digital counterpart, DTs can significantly contribute to the digitization of construction site management by actively automating waste management and resource distribution while mimicking actual site conditions, machinery, materials, and workforce behavior. Their application aids in developing highly accurate virtual models, improving intelligent construction methods, and allowing for the creation of components that minimize waste. Additionally, utilizing DTs fosters sustainable construction practices and decreases the production of superfluous materials [18]. A safety framework has been suggested that utilizes a DT platform to reduce possible risks at construction sites. This framework consistently merges sensor data with the DT to pinpoint high-risk zones, forecast accidents and their possible consequences, and notify workers of these threats [39]. DTs facilitate predictive maintenance and life cycle management, contributing to lower repair expenses, prolonging the durability of construction materials, enhancing resource utilization, and decreasing waste.

2.2.3 DT Applications in Post-construction Phase

The existing literature on post-construction DTs mainly emphasizes their application for real-time engagement, increased predictability, and better decision-making in operation and maintenance, along with the management of facilities and assets [33]. DTs empower facility managers to make essential choices regarding building operations, performance oversight, and optimization of energy use. By gathering real-time information, they improve operational effectiveness, aid in predictive maintenance, and enable well-informed decision-making [15]. DTs support facility managers in planning and assessing maintenance activities, as well as tracking and predicting the state of buildings. Furthermore, they provide building owners and managers with immediate insights into operational expenses, facilitating more knowledgeable and efficient decision-making [33].

During the operation and maintenance stage, oversight of the project typically transitions from the builder, making data management and accessibility more challenging. While the virtual model can accurately depict the asset [40], it often lacks a direct connection to the physical project that has been constructed [41]. The main use of DTs is in the management and upkeep of assets within the built environment. They are intended to improve decision-making based on data in asset management and operational processes [18]. Throughout the operation and maintenance stage, DTs allow operators to track energy usage and equipment conditions in real time, quickly detect and address problems, and enhance energy efficiency. Throughout the renovation and demolition stages, DT systems can aid construction teams in assessing recyclable materials, facilitating a more efficient evaluation of recycling needs and encouraging material conservation [37].

2.3 Simulations Enabled by DT Technology

Since the beginning of humanity, individuals have participated in simulations. A simulation is characterized by one process imitating another, and since all processes inherently progress over time, simulation is an essential aspect of human cognition. Historically, it has functioned as a means for strategizing, acquiring knowledge, and making choices. From ancient hunters sketching strategies in the soil to contemporary soldiers engaging in training drills, simulation has reliably assisted individuals in getting ready for actual actions. Throughout the ages, simulations have transformed from ancient Greek theater and medieval narratives into films and computer-generated mathematical models of the twentieth century. In the present day, advanced technology allows for remarkably intricate, photorealistic representations of actual objects and settings [26].

Modeling, simulation, and decision-making are closely linked and have emerged as crucial instruments in both science and engineering. In essence, a model encapsulates the features of an object or system, whereas a simulation illustrates its behavior [42]. High-fidelity modeling and simulation, driven by high-performance computing and sophisticated information processing, present a wide array of advantages. These benefits encompass the promotion of creativity and problem-solving capabilities [43], facilitation of experimentation and "what-if" scenarios, and provision of insights into limitations and causal relationships. They also aid in extensive data analysis and integration, potentially leading to groundbreaking scientific findings. Furthermore, they improve training and decision-support frameworks, lower expenses by embedding modeling directly into mission systems, streamline operations, and enable early testing and assessment of intricate systems. Additional benefits include reliable real-time control, optimized mission planning, trade-off analyses, predictive assessments, risk mitigation, accelerated design timelines, and reduced overall life cycle expenses [42]. DTs serve as virtual models of tangible objects, systems, or processes, enabling real-time simulation, observation, and assessment of their actual performance in the physical world [27].

DTs provide enhanced capabilities such as data evaluation, informed decision-making, and scenario modeling. The relationship between the physical and virtual entities is bidirectional: the connection from virtual to physical establishes a feedback loop, enabling the DT to affect the physical system through its analyses and choices, whereas the physical-to-virtual connection supplies the DT with real-time information from the real-world system [44]. According to [25], within the various categories of DTs, such as product DT and performance DT, production DT is focused explicitly on activities that occur prior to production. Its main roles encompass simulation, process validation, and analysis, all designed to enhance workflows ahead of actual manufacturing and to create more effective production techniques [27].

2.3.1 Data Integration and Modeling

The DT is exceptionally lifelike and capable of depicting one or several essential, interconnected vehicle systems, including propulsion and energy storage, avionics, life support, structural elements, and thermal management or thermal protection systems (TPS) [42]. In 2022, a theoretical modeling strategy was introduced that simplifies the intricate DT into more manageable models utilizing a 4C framework consisting of composition, context, component, and code [45]. These models are subsequently combined through information fusion, multi-scale association, and multi-context interaction [46].

Creating a DT involves building a virtual model of a physical entity, analyzing its data, and maintaining a synchronized connection to allow continuous monitoring. Sensors are integrated into the physical system to capture the necessary data for simulations with the virtual model, enabling an understanding of the object's behavior and supporting the development of appropriate solutions when required. The collected data is transmitted to the cloud via IoT technologies, where it is continuously synchronized with the virtual model to ensure that the representation reflects the most up-to-date state of the physical system [47]. DTs leverage the capabilities of the IoT to facilitate the transformation of raw sensor data into higher-level knowledge. Similarly, Rosen et al. [48] described a DT as a realistic representation of a process in its current state, capturing its behavior and interactions with the real-world environment, and it requires the existence of the corresponding physical component [49, 50].

2.3.2 Predictive Simulations

The DT serves numerous essential functions. It is capable of simulating upcoming missions prior to their launch to evaluate mission parameters, detect possible anomalies, and create strategies for mitigation, thereby aiding in certification. By incorporating environmental and system health updates, it can replicate real-time flight conditions for ongoing predictions [51]. Furthermore, it allows for in situ forensics to investigate faults or damages and can evaluate alterations to mission parameters—like malfunctioning components—to forecast structural effects, remaining lifespan, and best mitigation strategies, thus promoting informed decision-making [42]. This method facilitates the simulation of multi-scale, multidimensional, and multi-scenario environments within complex DTs [46].

A defining feature of prediction in DTs is their ability to forecast the future states of their physical counterparts, either through causal reasoning or probabilistic modeling. These so-called subspaces represented virtual domains in which predictive simulations could take place. The presence of multiple subspaces suggested that the number of possible simulations was essentially unlimited [7]. A DT evolves by adapting to its environment, a process made possible through the accurate simulation of data collected from embedded sensors. Achieving this requires expertise in simulation, AI, and ML [47]. Gabor et al. [52] described a DT as the simulation of a physical entity designed to predict its future states.

2.3.3 Operational Simulations

DT technology establishes a virtual environment in which production processes can be simulated and forecasted in parallel with real-world operations, thereby maximizing the utility of information and reducing the consumption of physical resources [53]. DT technology offers a range of capabilities, including continuous monitoring, real-time updates, scenario simulation, data analysis, operational control, outcome prediction, and performance optimization. During the design phase, DTs are capable of simulating the entire life cycle of a building and generating feedback that supports designers in evaluating performance under varying parameters. In the construction stage, DTs can detect and resolve potential problems, thereby improving quality and minimizing complexity. During the operation and maintenance phase, DTs continuously collect data from the physical environment, enabling precise diagnosis and analysis of the building's condition [54], streamlining maintenance, and allowing architects to store information for future projects [53, 55].

The continuous processing of data gathered from connected sensors enables real-time updates within the virtual model [47]. This stage of data mining and analysis is referred to as the interaction and convergence features of DTs. These features involve extracting data from the physical system and integrating both historical and real-time datasets to maintain a seamless synchronization channel between the physical and virtual environments [49]. For example, highway networks showcase the benefits of DT technology by utilizing information from roadside sensors and vehicles to facilitate real-time traffic observation and management. This approach helps in handling the intricacies of transportation networks and has attracted considerable interest from both the industrial sector and scholarly circles [56].

2.3.4 Collaboration and Visualization

DTs provide a platform for exploring and enhancing creativity in structural design, allowing multiple scenarios to be tested and potential outcomes visualized in realistic virtual environments. At the same time, organizational leaders can leverage these tools to implement more efficient management strategies, coordinate multidisciplinary teams, and better integrate social and operational resources throughout the project life cycle [53]. By integrating physical road infrastructure with virtual simulations, DTs facilitate continuous, two-way communication across all project phases. This includes design, construction, traffic management, and day-to-day operations, ensuring that information flows seamlessly between the real-world system and its digital counterpart [57–59].

DTs make it easier for people from different fields to work together. They provide interactive 3D environments that update in real time. Tools like Omniverse let designers, engineers, and managers check, change, and approve models together. This helps avoid mistakes and makes decisions faster. Clear and realistic visuals, like detailed renderings and immersive interfaces, help everyone understand complex systems. At the same time, organized workflows keep the digital models in sync with what actually happens, making teamwork smoother and more effective.

3 Omniverse

In recent times, there has been a significant rise in remote collaboration, as various teams often join forces to work on the same project [60]. This movement has increased the need for instruments that facilitate instantaneous collaboration, permitting users to collaborate from any place [1, 61]. Real-time collaboration plays a crucial role in the creation of 3D content, enabling artists, designers, and developers to collaboratively build intricate, dynamic virtual spaces. In contrast, conventional workflows typically operate in separate silos, hindering collaboration because of incompatible file formats and data structures and requiring manual data synchronization, which hampers productivity. To address these obstacles, Pixar unveiled the Universal Scene Description (USD) in 2016, establishing an open-source standard for the representation and exchange of 3D scenes. Expanding upon USD, NVIDIA introduced Omniverse in 2020, a development platform that integrates flawlessly with USD. Omniverse facilitates large-scale, intricate visualizations in real time.

USD surpasses traditional formats like OBJ, FBX, and Unity by functioning as a flexible interchange format and a strong foundation for scene assembly. Beyond handling geometry and textures, USD also accommodates shading, materials, rendering, animation, skinning, and physics, all while ensuring significant extensibility [2]. USD improves asset management by facilitating reuse across various scenes, minimizing storage needs, and enhancing pipeline efficiency. As a platform-agnostic solution, USD operates seamlessly on both Linux and Windows systems [62, 63]. Nonetheless, organizations may face challenges in adoption due to a lack of trained staff, as existing pipelines often need reconfiguration and users must learn new skills. The feature set is continually advancing, which requires attention to future compatibility, and certain functionalities from proprietary formats may still be absent. NVIDIA is actively advancing USD, focusing on enabling comprehensive Metaverse capabilities [64].

Figure 2 provides the ecosystem of the NVIDIA Omniverse platform, presenting its core functions, privileges, and ecosystem of tools. Omniverse supports real-time teamwork and also enables the integrated collaboration between designers, engineers, and experts through 3D design applications. In fact, Omniverse improves the speed of production cycles; additionally, it can provide secure and scalable access to assets by offering comprehensive enterprise support for increasing productivity and reducing downtime. The structure of Omniverse contains various elements such as APIs, services, and SDKs, integrating microservices. In addition, these elements demonstrate Omniverse as a powerful, scalable ecosystem for real-time, multiuser design, simulation, and digital collaboration.

3.1 Components of Omniverse

NVIDIA Omniverse is built on several core components that work together to support collaboration, simulation, and rendering within virtual environments [65]. A variety of Omniverse components are incorporated within NVIDIA Omniverse, including "Nucleus," "Connect," "Simulation," "Kit," and "RTX Renderer." A summary of each component, which is essential for the operation and smooth functioning of the project, is provided below [66].

- **Nucleus**: At the center of this ecosystem is Omniverse Nucleus, often described as the platform's collaboration engine. Acting as a database, it enables real-time asset sharing and modification across applications, services, and renderers, using a publish/subscribe model. It can be deployed on a private cloud, locally, or on a workstation, offering flexibility depending on user needs [65].
- **Connect**: Omniverse Connect serves as the link between Nucleus and digital content creation tools, relying on two-way connectors that ensure seamless live synchronization of assets and workflows across platforms [2].
- **Simulation**: NVIDIA's core physics simulation technologies provide developers with advanced, physically accurate, real-time world simulations. The main tools include NVIDIA PhysX, PhysX Flow, and PhysX Blast. PhysX handles high-performance rigid bodies, deformable objects, and vehicle dynamics, while PhysX Flow specializes in fire dynamics, combustible fluids, and smoke effects. PhysX Blast, on the other hand, is used for fracture and destruction simulations [66].
- **Kit**: Complementing this is Omniverse Kit, the software development kit that allows developers to build applications, extensions, and services, either through a customizable user interface or as headless microservices. For physically

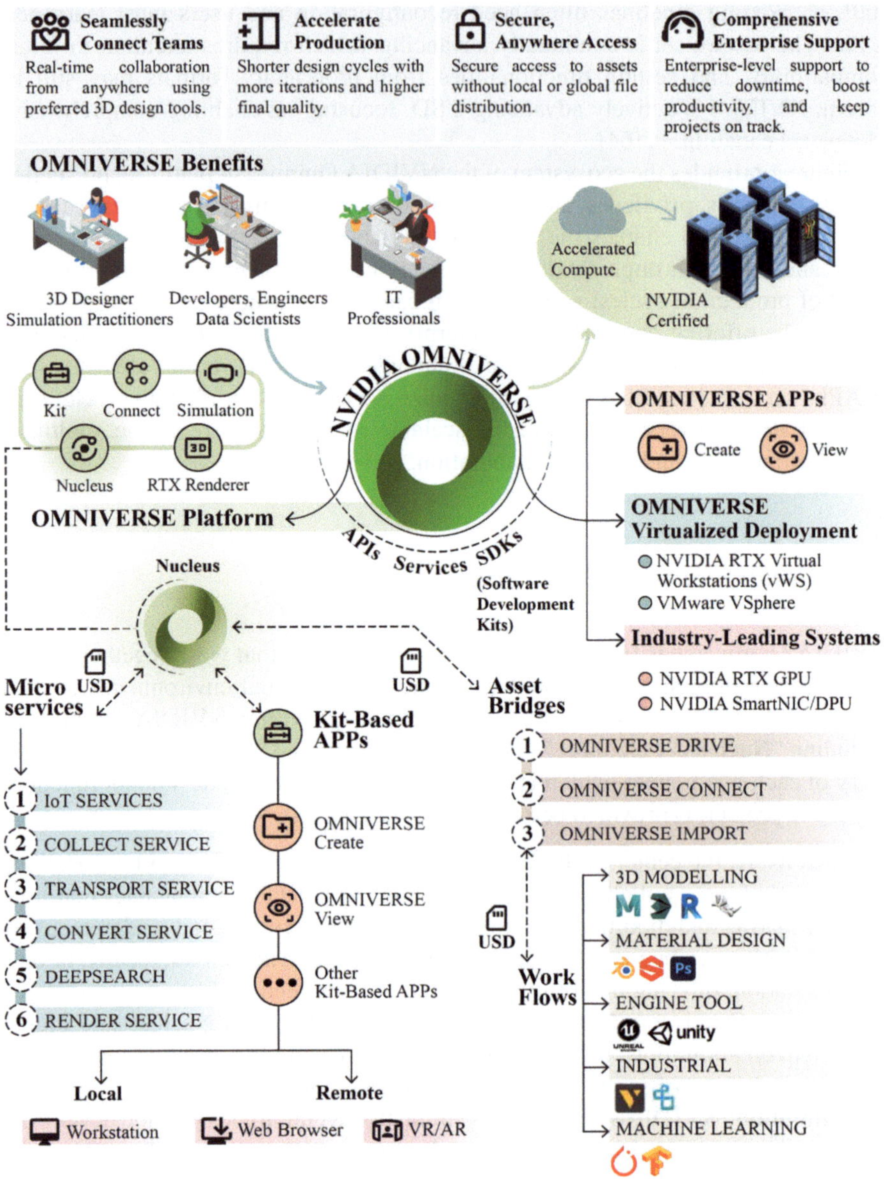

Fig. 2 Omniverse ecosystem

accurate real-time simulation, Omniverse Simulation integrates NVIDIA's physics technologies, including PhysX for rigid and soft-body dynamics; Flow for fluid, smoke, and fire effects; and Blast for destruction and fracture simulations.

- **RTX Renderer**: Finally, the Omniverse RTX Renderer offers a scalable multi-GPU rendering solution that supports both real-time ray tracing for interactive performance and path tracing for cinematic-quality visuals while also maintaining compatibility with Pixar's Hydra rendering architecture [67].

3.2 *Transforming the AEC Industry with Omniverse*

For the first time, teams spread across the globe can work together in real time on a unified model and refine designs, even when utilizing various software applications. This capability is enabled by NVIDIA Omniverse Connectors, which integrate top industry tools into a common collaborative environment [60, 65]. Consequently, designers and engineers are able to utilize the applications that are most appropriate for their specific tasks while effortlessly collaborating with their peers [2, 65, 68]. NVIDIA Omniverse facilitates effortless collaboration by bringing together project teams on a unified interactive platform, permitting real-time creation of architectural models, even when team members utilize various software applications [60]. With just one click, teams can create physically precise, visually impressive ray-traced designs without the need for extensive data preparation or model simplification. Omniverse also speeds up the approval process by facilitating quick iterations and idea exploration while delivering RTX ray-traced quality exports. High-fidelity models can be shared and accessed by teams, clients, and contractors on any device, anywhere, improving efficiency and communication [69].

By utilizing Omniverse Connectors, architects and designers are able to operate within their chosen software, including Autodesk Revit, McNeel & Associates Rhino, and Trimble SketchUp, all while collaborating on a unified scene file in Omniverse Nucleus [65, 68]. This real-time connection enables users to efficiently refine a project in Omniverse Create, which visualizes the scene instantaneously with realistic materials, lighting, reflections, and shadows. Meanwhile, project reviewers and collaborators can immediately observe modifications through Omniverse View [69]. Omniverse Create, an application designed on the NVIDIA Omniverse Kit™, facilitates sophisticated scene composition by enabling users to interactively construct, illuminate, simulate, and render USD scenes in real time [62]. It interfaces with Omniverse Nucleus and integrates effortlessly with prominent design software such as Autodesk 3ds Max, Revit, and Trimble SketchUp, offering a highly precise, shared viewport for extensive projects.

This application is adept at managing enormous scenes with live-linked assets, supporting scalable performance across numerous GPUs or nodes, and providing real-time, ray-traced outcomes while allowing designers to apply physically accurate materials, environments, and lighting with immediate updates. At the heart of Omniverse Create lies the Omniverse RTX Renderer, which utilizes RT Cores for ray tracing and Tensor Cores for AI-driven denoising and upscaling. With the capability to alternate between real-time RTX ray-traced and cinematic-quality path-traced modes, teams can render intricate scenes with both high fidelity and interactivity.

In addition to its technical rendering capabilities, Omniverse Create elevates realism through an integrated vegetation library that can manage billions of polygons, supports Illuminating Engineering Society (IES) lighting for precise simulation of actual fixtures, and features a Sun Study tool that takes into account geographic location and dynamic clouds for realistic lighting assessments. Furthermore, the incorporation of extended reality (XR) through NVIDIA CloudXR™ facilitates immersive engagement with projects on head-mounted displays, smartphones, or tablets, enabling collaborative exploration within photorealistic, real-time settings. Omniverse incorporates cutting-edge technologies like real-time RTX™ rendering, lighting simulations, physics, and various other simulation tools, enhancing both innovation and productivity. By merging these functionalities, Omniverse empowers designers, architects, and engineers to develop, visualize, and assess projects more rapidly than ever before, facilitating collaboration from any location globally and providing unmatched advantages through optimized workflows, seamless teamwork, accelerated design processes, and impressive outcomes [2, 65, 68, 69].

4 Integrating DTs with Omniverse

The physical tier encompasses production methods in which humans and cyber-physical systems (CPS) work together, whereas the digital tier emphasizes data analysis aimed at enhancing decision-making. The connection between these two levels is facilitated by human-machine interfaces (HMIs) and CPS [70, 71]. Conversely, in the industrial virtual realm, the tangible layer is entirely substituted by a virtual equivalent, where assets are present exclusively within a carefully designed computer-generated setting [72]. Avatars symbolize human beings, employing inverse reality methods to interpret their environment; make educated choices; and engage with other avatars. In this virtual setting, cyber-physical systems (CPS) are virtualized via DTs, serving as crucial links between the virtual and digital spheres, utilizing virtual sensing, integrated intelligence, and uninterrupted connectivity across various applications [71, 73].

The digital aspect of this industrial virtual setting is embodied by the Metaverse and the Omniverse. The Metaverse mainly facilitates efficient and interoperable interactions among DTs and oversees their incorporation with a broader array of assets within the virtual realm. In contrast, the Omniverse operates on a more advanced level, outlining the intricate setups for each cyber-physical system (CPS) along with the particular parameters of the virtual environment [71].

Creating a DT within the Omniverse enables architects, engineers, and construction teams to work together in evaluating designs, facilitating faster development, and ensuring that projects remain on track. DT technology facilitates the execution of numerous tasks within a virtual setting, enabling simulations to pinpoint and mitigate bottlenecks and various challenges. For the implementation of edge computing and AI applications, including real-time product recognition, precise data, operational insights, and auxiliary technologies are crucial for a network of intelligent factories and distribution centers [74–77]. This results in quicker and more adaptable evaluations of products and completion of orders.

DTs are extensively utilized for the training and assessment of robotic systems. Training supervised ML models requires a significant amount of data. DTs can generate a highly accurate virtual setting that yields substantial volumes of test data [78], facilitating the efficient training of ML models, which can subsequently be applied to the physical system or robot. NVIDIA has demonstrated this ability with its newly introduced NVIDIA Omniverse platform, partnering with companies such as Pepsi and Amazon to create DTs of warehouses and distribution hubs for the training of AI models [70, 73, 79].

4.1 Evolution of Research Themes on Omniverse and DTs in Construction

Figure 3 depicts the evolution of research themes based on the combination of Omniverse as a collaborative 3D virtual world and DTs as a digital mirror of the real world. These two ideas are located at the center of this connection, presenting how closely they connect to other trending topics like AI, XR, Industry 5.0, and cloud computing. Additionally, Omniverse is a new idea of a shared digital space where different virtual worlds, simulations, and DTs can work together seamlessly. The idea of the Omniverse has been emerging and gradually gaining attention since around 2023, as more people and organizations have started exploring its potential and implications. When combined with DTs, the Omniverse provides a strong

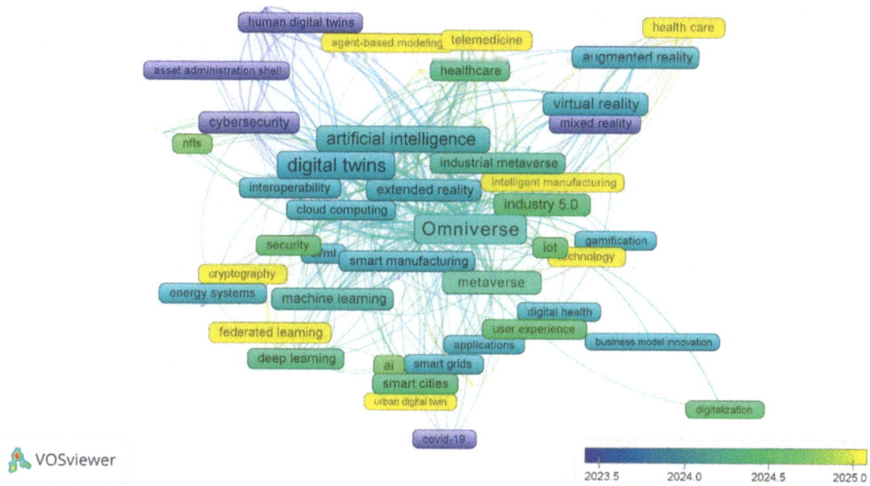

Fig. 3 Mapping the relationship between Omniverse and DTs

virtual space where real-world systems can be copied, tested, and improved, making it easier for people to collaborate and innovate on a larger scale.

The color scale starts with purple in 2023, where terms like human DTs and mixed reality (MR) appear. This reflects early interests in the building industry around integrating immersive technologies and simulating human interactions within the digital construction environment. Moving into the blue-green stage around 2024, concepts such as machine learning, cloud computing, and smart manufacturing are being emphasized. In construction, AI is being used to optimize structural design, simulate energy performance, and enable collaborative building information modeling (BIM) through cloud-based platforms. In the light green area, keywords such as digital health, smart cities, and Industry 5.0 show the expansion of these technologies into human-centered construction practices. As an example, Industry 5.0 in the building sector emphasizes collaboration between humans and machines, where robots handle repetitive construction tasks while workers focus on creative design and sustainable solutions. Finally, the yellow area, which is displayed in 2025, presents new topics like federated learning, intelligent manufacturing, and telemedicine, which in the construction field translate to decentralized AI learning across multiple sites, smart factories producing prefabricated components, and designing buildings that integrate digital health and remote care spaces for future urban living.

The Omniverse is a collaborative 3D virtual world connected to the emerging metaverse, and it is increasingly being used together with DTs, which are accurate digital copies of real-world systems, to support innovation in the building industry. This combination makes it possible to run immersive simulations, optimize processes with AI, and coordinate work through cloud-based platforms while connecting with trends such as human DTs, MR, Industry 5.0, and smart manufacturing. By offering a shared virtual space where real systems can be modeled, tested, and improved, the Omniverse and DTs facilitate collaboration between humans and machines, promote sustainable design, and expand construction practices into human-centered areas like digital health and smart cities, thus shaping the future of construction.

4.2 Leveraging Omniverse for DT-Based Simulations

Digital assessments carried out in the DT environment generate significant data that aids in evaluating the effectiveness of new products or manufacturing methods across different scenarios. Nonetheless, DTs frequently do not provide an interface that enables individuals to engage with the virtual environment as intuitively as they do with the physical world [71]. Consequently, the degree of engagement between individuals and the digital environment in industrial contexts continues to be restricted [80]. To fill this void, applications are being created that utilize immersive technologies like VR and AR, allowing individuals to engage with the digital actions of cyber-physical systems in a more lifelike manner.

Nonetheless, the majority of existing immersive reality applications that utilize DTs concentrate on examining human engagement with cyber-physical systems mainly for educational objectives, training, prototyping, or examining dynamic environments, but the cooperative industrial landscape typical of Industry 4.0 includes far more than merely interactions between humans and machines [81]. It similarly illustrates the merging of various industrial resources. Consequently, a more comprehensive strategy is essential, in which the interactions among individuals and other industrial assets within the virtual realm closely mimic the natural interactions found in the physical world, thereby fostering a more engaging and authentic virtual experience [82, 83].

It is also widely recognized that the productivity of industries relies on efficient collaboration among operators [84, 85]. Nonetheless, interactions between individuals in industrial settings have not been thoroughly examined, especially concerning cooperative activities within virtual environments. The significance of virtual human-to-human engagement transcends mere technology, closely aligning with the emergence of the forthcoming industrial framework, Industry 5.0, which prioritizes a human-centered methodology. In addition to its technological influence, this transformation holds considerable social consequences, fostering general welfare within a robust ecosystem [86]. This enduring, multiuser digital landscape, often referred to as the Metaverse, signifies a cohesive domain where the distinctions between the real and virtual realms fade, establishing a novel arena for human engagement [71].

The Metaverse is intricately linked to cyberspace and illustrates the ways in which individuals engage with both the physical and virtual realms [87]. The Metaverse is described as an all-encompassing digital realm where various worlds can converge, a notion frequently termed the Omniverse. Omniverse is a platform centered on extensions and microservices, exemplified by Omniverse Create, now referred to as Omniverse USD Composer. It features 300 Python-script extensions that offer nearly total customization, along with a microservice architecture enabling developers to craft custom extensions or develop new applications utilizing only the necessary components [2, 88].

The Metaverse enhances efficiency throughout different stages of the product life cycle by facilitating numerous precise simulations at a reduced cost, eliminating the need for physical samples. Through the utilization of DTs, the Metaverse boosts operational effectiveness, elevates product quality, reduces quality-related risks, and facilitates real-time engagement between customers and developers to refine product life cycle design throughout the development process. A notable illustration of the industrial Metaverse is NVIDIA Omniverse™. The industrial Metaverse gathers information from multiple sensors and operational processes, delivering practical analytics and insights for decision-making aimed at enhancing production efficiency, lowering expenses, and optimizing sales value [87]. Additionally, through the implementation of federated learning, AI, and BCT technology, Industrial IoT nodes can enhance trust, decentralization, and privacy, all while improving the efficiency of data sharing.

NVIDIA Omniverse is an adaptable, multi-GPU, open-source framework that facilitates worldwide cooperation among creators, artists, and enterprises [88]. It offers an extensive array of development resources and excels in real-time 3D design and simulation technologies. Utilizing NVIDIA's proficiency in GPU technology, Omniverse incorporates AI, blockchain, and NVIDIA RTX™ features to improve interactive graphics and computational efficiency [64]. The platform caters to a diverse audience, encompassing researchers, engineers, and digital designers, and employs Pixar's USD to facilitate the development of applications, extensions, and microservices. Its modular design permits these advancements to be exchanged and incorporated across various projects, fostering adaptability, teamwork, and expandability in workflows for 3D content creation [67].

4.3 Omniverse in Action: Real-Time Collaboration and Industrial DTs

Introduced in 2019, NVIDIA Omniverse has revolutionized workflows in media and entertainment by facilitating effortless collaboration within 3D production pipelines. In the past, artists depended on distinct tools and faced lengthy procedures involving data transfer, format changes, and importing before they could collectively visualize assets. Omniverse addresses these challenges by offering real-time synchronization of geometry, materials, and other resources via a central server, which can be accessed locally or through the cloud. This setup allows artists to continue utilizing their preferred software while ensuring that updates are immediately reflected across all interconnected applications. By integrating with both content creation platforms and rendering engines for games or films, Omniverse eliminates the need for manual conversions, thereby speeding up production and enhancing collaborative creativity [89].

To facilitate this, assets and scenes are depicted through an enhanced version of Pixar's USD [90]. Generally, assets are kept and exchanged in binary or compressed formats, with modifications sent to Omniverse via incremental updates. Additionally, the platform accommodates material descriptions formulated in the material definition language (MDL) for physically based materials [65]. The Omniverse features a reference application for viewing and editing, known as Omniverse Kit, which utilizes cutting-edge GPU capabilities to provide exceptional visuals with optimal performance [89].

BMW has collaborated with NVIDIA to bring its concept of the "factory of the future" to life by utilizing NVIDIA Omniverse and AI technologies. This partnership has led to the development of a DT of its Regensburg factory, which was initially presented at the NVIDIA GTC conference in November 2021 [64, 91]. Omniverse provides a platform for simulation and collaboration, enabling the integration of current workflows into collective virtual settings. Through this virtual factory, BMW is able to replicate its comprehensive planning and manufacturing processes, visualizing operations from inception to completion, while AI-powered worker avatars engage with robots and manage components, enhancing productivity and ergonomics within the simulated production line [91].

4.4 Integrating DTs and Omniverse for Construction 5.0

With digital enablement and the shift toward digital control, the anticipated transition to Industry 5.0 highlights the partnership between human and machine intelligence, with the goal of facilitating more informed decision-making and creating more integrated, sustainable systems. Furthermore, the merging of the physical and digital domains is progressing beyond current limitations. In the construction sector, digital twins enhanced by NVIDIA Omniverse facilitate lifelike simulations of design choices, encompassing aspects such as building location, materials, and lighting, through AI-driven evaluations. These digital twin solutions can effortlessly connect with popular 3D applications like Autodesk Maya, 3ds Max, and Unreal Engine via connectors that allow for two-way data transfer. This compatibility enables architects, engineers, and designers to utilize their preferred software while ensuring that models remain synchronized in Omniverse, thus removing the necessity for manual updates. Additionally, the platform bolsters interdisciplinary collaboration and produces photorealistic digital twins using RTX real-time ray tracing, resulting in highly precise virtual representations of real-world environments.

Omniverse supports real-time collaboration, simulation, and visualization in large-scale construction projects, reducing errors, accelerating timelines, and improving budget control. Bentley Systems integrates Omniverse with its iTwin platform to create and manage digital twins, which enhance planning, design, construction, maintenance, and scenario modeling across an asset's life cycle. Digital twins significantly boost efficiency and productivity in the construction industry. Meanwhile, research into machine learning and computer vision is needed to enable safe, effective human-robot collaboration on-sites. Construction 5.0, aligned with Industry 5.0, should emphasize human-centered, sustainable innovations tailored to the sector, though interactive robots remain largely experimental [92–94].

Figure 4 shows the layered framework of applying DTs and Omniverse in construction, from technologies to industry-specific applications and benefits. The first layer highlights the integration of DTs and Omniverse as the foundation of real-time collaboration in construction. DTs can provide the data of assets and also enable simulation and support life cycle monitoring of buildings and infrastructure. On the other side, Omniverse ensures interoperability between different tools by GPU-powered visualization, and seamless collaboration among stakeholders. Together, they create a digital foundation where design, planning, and operations can be connected in a unified environment.

The second layer demonstrates how DTs and Omniverse apply across the building life cycle. In planning and design, their integration with BIM and 3D visualization enables teams to refine models and collaborate effectively.

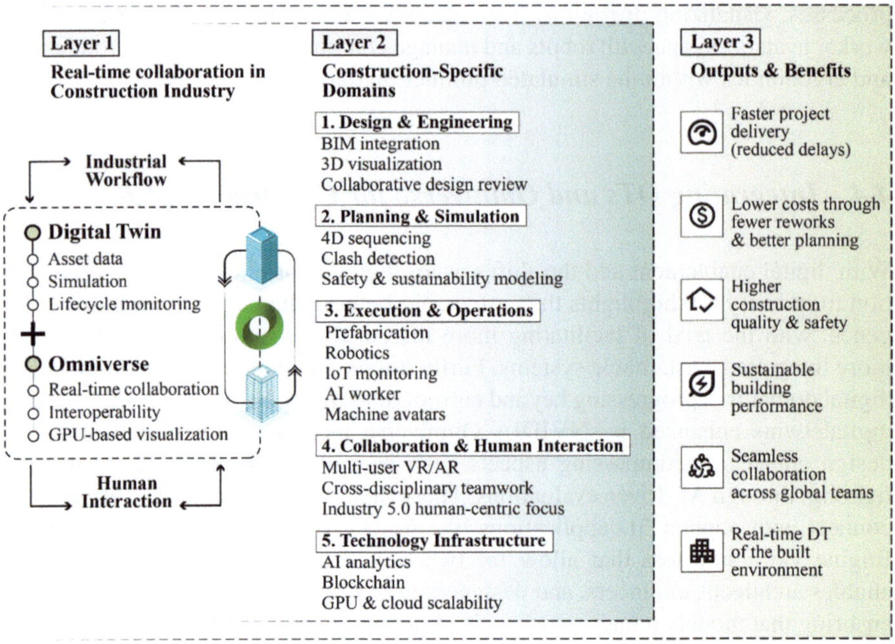

Fig. 4 Integration of Omniverse and DT in the construction industry

Throughout construction, DTs offer real-time insights into task sequencing, clash management, prefabrication processes, robotics, IoT devices, and AI-supported worker activities, enhancing efficiency on site. In the operation and maintenance phase, they track building performance to support predictive maintenance, energy optimization, and VR/AR-based collaboration aligned with Industry 5.0. Even in demolition, DTs supply data for safer and more sustainable practices. Across all these stages, AI analytics, blockchain security, and scalable GPU-cloud resources form the technological foundation that makes this ecosystem reliable and actionable.

The third layer shows the real benefits of this integrated approach. Projects can be finished faster with fewer interruptions, and costs are reduced thanks to better planning and less rework. Quality and safety improve because risks are managed more effectively, and sustainability goals are supported through performance modeling. Beyond the technical benefits, this integration also makes collaboration easier for international teams and provides real-time DTs of the built environment, offering continuous insights throughout the project.

5 Conclusion

The combination of DT and NVIDIA Omniverse technology creates valuable applications and opportunities in multiple industries, such as the construction industry, which is transforming design, implementation, and life cycle management. DTs develop an interactive, all-encompassing digital representation of a tangible asset by merging real-time information with simulation. This idea represents a real-time operational duplicate that mirrors the current condition of the asset or system. When used with Omniverse, these features help create collaborative virtual spaces that keep models consistent, improve work processes, and allow teams from different fields and locations to make smart decisions in real digital environments. Besides making operations more efficient, the DT-Omniverse system also supports sustainability by saving energy, cutting down waste, and using predictive maintenance. It also promotes a design method that focuses on people, which matches the goals of Industry 5.0. Examples from architecture, engineering, and manufacturing show how Omniverse improves digital processes by using realistic visualizations, interactive XR tools, and AI-powered design. Even though there are challenges like setting standards, preparing workers, and managing data, the teamwork between DTs and Omniverse helps create smarter, connected, and adaptable industries. This combination represents more than just a new technology—it marks a big change, helping construction and related areas move toward more teamwork, stronger systems, and better sustainability.

AI Use Disclosure Statement The authors did not use any generative AI or AI-assisted technologies in the creation of this manuscript.

References

1. DeFilippis E, Impink SM, Singell M, Polzer JT, Sadun R (2020) Collaborating during coronavirus: the impact of COVID-19 on the nature of work. National Bureau of Economic Research
2. Vatanen J (2024) Exploring NVIDIA Omniverse ecosystem
3. Golabchi M, Faraji A (2010) Project strategic management. Press University of Tehran, Tehran, Iran
4. Faraji A, Rashidi M, Meydani Haji Agha T, Rahnamayiezekavat P, Samali B (2022) Quality management framework for housing construction in a design-build project delivery system: a BIM-UAV approach. Buildings 12(5):554
5. Rahnamayiezekavat P, Sorooshnia E, Rashidi M, Faraji A, Mostafa S, Moon S (2022) Forensic analysis of the disputes typology of the NSW construction industry using PLS-SEM and prospective trend analysis. Buildings 12(10):1571
6. Mehrinejad Khotbehsara M, Zadshir M, Mehdizadeh Miyandehi B, Mohseni E, Rahmannia S, Fathi S (2014) Rheological, mechanical and durability properties of self-compacting mortar containing nano-TiO_2 and fly ash. J Am Sci 10:222–228
7. Grieves M, Hua EY (2024) Defining, exploring, and simulating the digital twin Metaverses. In: Digital twins, simulation, and the Metaverse: driving efficiency and effectiveness in the physical world through simulation in the virtual worlds. Springer, pp 1–31

8. Faraji A, Homayoun Arya S (2023) Proposing an integrated time-cost management model based on building information modeling (BIM) and blockchain technology (BCT) smart contract development approach in the construction industry. J Model Eng 21(74):191–206
9. Gh SH, Miyandehi BM, Khotbehsara MM, Tarbiat M (2015) Study quality steel mill slag for use in concrete containing metakaolin. In: International conference on advances in engineering, Tehran, Iran
10. Mystakidis S (2022) Metaverse. Encyclopedia 2(1):486–497
11. Bansal G, Rajgopal K, Chamola V, Xiong Z, Niyato D (2022) Healthcare in Metaverse: a survey on current Metaverse applications in healthcare. IEEE Access 10:119914–119946
12. López-Díez J (2021) Metaverse: Year One. Mark Zuckerberg's video keynote on Meta (October 2021) in the context of previous and prospective studies on Metaverses. Pensar Public 15(2):299–303
13. Zhang J, Cheng JCP, Chen W, Chen K (2022) Digital twins for construction sites: concepts, LoD definition, and applications. J Manag Eng 38(2). https://doi.org/10.1061/(asce)me.1943-5479.0000948
14. Faraji A, Arya SH, Ghasemi E, Soleimani H (2023) Considering the digital twin as the evolved level of BIM in the context of construction 4.0. In: Proceedings of the 1st national & international conference on architecture, advanced technologies and construction management (IAAC), Tehran, Iran
15. Opoku D-GJ, Perera S, Osei-Kyei R, Rashidi M (2021) Digital twin application in the construction industry: a literature review. J Build Eng 40:102726
16. Ellul C, Hamilton N, Pieri A, Floros G (2024) Exploring data for construction digital twins: building health and safety and progress monitoring twins using the unreal gaming engine. Buildings 14(7). https://doi.org/10.3390/buildings14072216
17. Movagharnejad K, Zareei F, Mehdizadeh B, Salahi S, Lashkenari MS (2015) Prediction of the surface tension of hydrocarbons. Pet Sci Technol 33(9):1008–1016
18. Zribi H, Ben Abid T, Elloumi A, Hani Y, Bechir Graba B, Elmhamedi A (2025) Industry 4.0: digital twins characteristics, applications, and challenges in-built environments. Prod Manuf Res 13(1). https://doi.org/10.1080/21693277.2025.2456277
19. Mousavi Y, Gharineiat Z, Karimi AA, McDougall K, Rossi A, Gonizzi Barsanti S (2024) Digital twin technology in built environment: a review of applications, capabilities and challenges. Smart Cities. https://doi.org/10.3390/smartcities7050101
20. Singh M, Fuenmayor E, Hinchy EP, Qiao Y, Murray N, Devine D (2021) Digital twin: origin to future. Appl Syst Innov. https://doi.org/10.3390/asi4020036
21. Gartner INC (2015) Gartner identifies the top 10 strategic technology trends for 2013
22. Qi Q, Tao F, Zuo Y, Zhao D (2018) Digital twin service towards smart manufacturing. Procedia CIRP 72:237–242
23. Marr B (2020) These 25 technology trends will define the next decade. Forbes:1–7
24. Madni AM, Madni CC, Lucero SD (2019) Leveraging digital twin technology in model-based systems engineering. Systems 7(1):7
25. Grieves M, Vickers J (2016) Digital twin: mitigating unpredictable, undesirable emergent behavior in complex systems. In: Transdisciplinary perspectives on complex systems: new findings and approaches. Springer, pp 85–113
26. Grieves M, Hua EY (2024) Defining, exploring, and simulating the digital twin Metaverses. Springer, pp 1–31. https://doi.org/10.1007/978-3-031-69107-2_1
27. Iliuţă ME, Moisescu MA, Pop E, Ionita AD, Caramihai SI, Mitulescu TC (2024) Digital twin—a review of the evolution from concept to technology and its analytical perspectives on applications in various fields. Appl Sci (Switz) 14(13). https://doi.org/10.3390/app14135454
28. Abdelrahman M, Macatulad E, Lei B, Quintana M, Miller C, Biljecki F (2025) What is a Digital Twin anyway? Deriving the definition for the built environment from over 15,000 scientific publications. Build Environ 274. https://doi.org/10.1016/j.buildenv.2025.112748
29. Faraji A, Arya SH, Ghasemi E, Soleimani H (2024) IoT core applications in AEC: investigation of the most compatible technologies of Industry 4.0. ResearchGate, Berlin, Germany

30. Mehdizadeh B, Vessalas K, Ben B, Castel A, Deilami S, Asadi H (2023) Advances in characterization of carbonation behavior in slag-based concrete using nanotomography. In: Nanotechnol Constr Circ Econ (NICOM 2022), Melbourne, pp 297–308
31. Faraji A, Arya SH, Ghasemi E (2023) Investigation of the prospective developments in the architectural design process in the fourth industrial revolution
32. Faraji A et al (2024) A conceptual framework of decentralized blockchain integrated system based on building information modeling to steering digital administration of disputes in the IPD contracts. Constr Innov 24(1):384–406
33. Adu-Amankwa NAN, Pour Rahimian F, Dawood N, Park C (2023) Digital Twins and Blockchain technologies for building lifecycle management. Autom Constr. https://doi.org/10.1016/j.autcon.2023.105064
34. Mehdizadeh Miyandehi B, Vessalas K, Castel A, Mortazavi M (2023) Investigation of carbonation behaviour in high-volume GGBFS concrete for rigid road pavements. ASCP (Australian Society for Concrete Pavements)
35. Häkkinen T, Kuittinen M, Ruuska A, Jung N (2015) Reducing embodied carbon during the design process of buildings. J Build Eng 4:1–13
36. Doumbouya L, Gao G, Guan C (2016) Adoption of the Building Information Modeling (BIM) for construction project effectiveness: the review of BIM benefits. Am J Civ Eng Archit 4(3):74–79
37. Zhang Z et al (2024) A review of digital twin technologies for enhanced sustainability in the construction industry. Buildings. https://doi.org/10.3390/buildings14041113
38. Faraji A, Arya SH, Ghasemi E, Shiri H (2024) The role of artificial intelligence (AI) in advancing construction 4.0 practices
39. Chronopoulos C, Johansen KW, Teizer J, Schultz C, Esterle L (2021) Towards a holistic, self-organised safety framework for construction. In: 2021 IEEE international conference on autonomic computing and self-organizing systems companion (ACSOS-C). IEEE, pp 238–243
40. Mehdizadeh B, Vessalas K, Castel A, Asadi H, Ben B, Deilami S (2024) Correlation behaviours for accelerated and natural carbonation testing of fly ash and ground granulated blast-furnace slag-based concrete mixes. Concr Aust 50:30–38
41. Anderl R, Haag S, Schützer K, Zancul E (2018) Digital twin technology—an approach for Industrie 4.0 vertical and horizontal lifecycle integration. it-Inf Technol 60(3):125–132
42. Shafto M, Rich MC, Glaessgen DE, Kemp C, Lemoigne J, Wang L (2010) DRAFT MoDeling, SiMulATion, inFoRMATion technology & PRoceSSing RoADMAP technology Area 11. NASA
43. Soltaninejad M, Noorzai E, Faraji A (2025) Enhancing the level of confidence in the safe fire evacuation operation employing a combination of BIM and fuzzy TOPSIS method. Smart Sustain Built Environ 14(1):163–189
44. Kreuzer T, Papapetrou P, Zdravkovic J (2024) Artificial intelligence in digital twins—a systematic literature review. Data Knowl Eng 151:102304
45. Jia W, Wang W, Zhang Z (2022) From simple digital twin to complex digital twin Part I: A novel modeling method for multi-scale and multi-scenario digital twin. Adv Eng Inform 53:101706
46. Latsou C, Ariansyah D, Salome L, Ahmet Erkoyuncu J, Sibson J, Dunville J (2024) A unified framework for digital twin development in manufacturing. Adv Eng Inform 62. https://doi.org/10.1016/j.aei.2024.102567
47. Rampini L, Re Cecconi F (2022) Artificial intelligence in construction asset management: a review of present status, challenges and future opportunities. J Inf Technol Constr 27:884–913. https://doi.org/10.36680/j.itcon.2022.043
48. Rosen R, Von Wichert G, Lo G, Bettenhausen KD (2015) About the importance of autonomy and digital twins for the future of manufacturing. IFAC-papersonline 48(3):567–572
49. Tao F et al (2019) Digital twin-driven product design framework. Int J Prod Res 57(12):3935–3953

50. Opoku D-GJ, Perera S, Osei-Kyei R, Rashidi M, Bamdad K, Famakinwa T (2023) Barriers to the adoption of digital twin in the construction industry: a literature review. Informatics 10:14
51. Eftekhar Afzali S, Ghasemi M, Rahimiratki A, Mehdizadeh B, Yousefieh N, Asgharnia M (2025) Compaction and compression behavior of waste materials and fiber-reinforced cement-treated sand. J Struct Des Constr Pract 30(2):4025007
52. Gabor T, Belzner L, Kiermeier M, Beck MT, Neitz A (2016) A simulation-based architecture for smart cyber-physical systems. In: 2016 IEEE international conference on autonomic computing (ICAC). IEEE, pp 374–379
53. Nguyen TD, Adhikari S (2023) The role of BIM in integrating digital twin in building construction: a literature review. Sustainability 15(13):10462
54. Faraji A, Meydani Haji Agha T, Homayoon Arya S, Ghasemi E, Rahnamayiezekavat P (2024) Hybrid study of quantitative–qualitative analysis to recognize the most cost-effectiveness criteria to develop affordable mass housing. Buildings 14(4):1057
55. Qi Q, Tao F (2018) Digital twin and big data towards smart manufacturing and Industry 4.0: 360 degree comparison. IEEE Access 6:3585–3593
56. Kušić K, Schumann R, Ivanjko E (2023) A digital twin in transportation: real-time synergy of traffic data streams and simulation for virtualizing motorway dynamics. Adv Eng Inform 55. https://doi.org/10.1016/j.aei.2022.101858
57. Liao X et al (2021) Cooperative ramp merging design and field implementation: a digital twin approach based on vehicle-to-cloud communication. IEEE Trans Intell Transp Syst 23(5):4490–4500
58. Dasgupta S, Rahman M, Lidbe AD, Lu W, Jones S (2021) A transportation digital-twin approach for adaptive traffic control systems. arXiv preprint arXiv:2109.10863
59. Argota Sánchez-Vaquerizo J (2022) Getting real: the challenge of building and validating a large-scale digital twin of Barcelona's traffic with empirical data. ISPRS Int J Geoinf 11(1):24
60. Balin S, Bolognesi CM, Borin P (2025) Immersive representation exercise in BIM through the integration of prerequisites and requirements: structuring of an artefact applied to a case study. In: Procedia computer science. Elsevier B.V, pp 155–166. https://doi.org/10.1016/j.procs.2025.01.079
61. Neeli H, Tran KQ, Velazco-Garcia JD, Tsekos NV (2024) A multiuser, multisite, and platform-independent on-the-cloud framework for interactive immersion in holographic XR. Appl Sci (Switz) 14(5). https://doi.org/10.3390/app14052070
62. Ahn S, Lee S, Hwang L, Kwon S (2025) Studying the Universal Scene Description (USD) file format from a Digital Twin Convergence Perspective. Int J Adv Smart Converg:224–241
63. Chamusca IL, Ferreira CV, Murari TB, Apolinario AL, Winkler I (2023) Towards sustainable virtual reality: gathering design guidelines for intuitive authoring tools. Sustainability. https://doi.org/10.3390/su15042924
64. Ahmed N, Afyouni I, Dabool H, Al Aghbari Z (2024) A systemic survey of the Omniverse platform and its applications in data generation, simulation and Metaverse. Front Comput Sci. https://doi.org/10.3389/fcomp.2024.1423129
65. Smith H (2025) The Metaverse-industrial complex. Inf Commun Soc 28(5):797–814. https://doi.org/10.1080/1369118X.2024.2423346
66. What is NVIDIA Omniverse?—Explained at Moralis Academy. https://academy.moralis.io/blog/nvidia-omniverse-explained. Accessed 12 Aug 2024
67. Omniverse Platform. https://www.nvidia.com/en-us/omniverse/. Accessed 30 Aug 2025
68. NVIDIA OMNIVERSE FOR ARCHITECTURE, ENGINEERING, AND
69. NVIDIA OMNIVERSE FOR AEC (2021)
70. Guerra-Zubiaga DA, Aksu M, Richards G, Kuts V (2025) Integrating Digital Twin software solutions with collaborative industrial systems: a comprehensive review for operational efficiency. Appl Sci. https://doi.org/10.3390/app15137049
71. Martínez-Gutiérrez A, Díez-González J, Perez H, Araújo M (2024) Towards Industry 5.0 through Metaverse. Robot Comput Integr Manuf 89:102764
72. Faraji A, Homayoon Arya S, Ghasemi E, Soleimani H, Rahnamayiezekavat P (2023) A constructability assessment model based on BIM in urban renewal projects in limited lands. Buildings 13(10):2599

73. Bhatia N et al (2025) Visualizing digital twins of fusion power plants using NVIDIA Omniverse. AIP Adv 15(4). https://doi.org/10.1063/5.0261883
74. Faraji A, Hosseini A, Kermani MZ, Mashatan N, Ardestani S (2023) Vertical farming: vertical farming; an innovative agricultural method to the urban and environmentally sustainable development. J Innov Sustain RISUS 14(3):166–181
75. Agnusdei GP, Elia V, Gnoni MG (2021) A classification proposal of digital twin applications in the safety domain. Comput Ind Eng 154:107137
76. Kritzinger W, Karner M, Traar G, Henjes J, Sihn W (2018) Digital Twin in manufacturing: a categorical literature review and classification. IFAC-PapersOnline 51(11):1016–1022
77. Gallala A, Kumar AA, Hichri B, Plapper P (2022) Digital Twin for human–robot interactions by means of Industry 4.0 Enabling Technologies. Sensors 22(13):4950
78. Faraji A, Rashidi M, Rezaei F, Rahnamayiezekavat P (2023) A meta-synthesis review of occupant comfort assessment in buildings (2002–2022). Sustainability 15(5):4303
79. Asad U, Khan M, Khalid A, Lughmani WA (2023) Human-centric digital twins in industry: a comprehensive review of enabling technologies and implementation strategies. Sensors 23(8):3938
80. Dincelli E, Yayla A (2022) Immersive virtual reality in the age of the Metaverse: a hybrid-narrative review based on the technology affordance perspective. J Strateg Inf Syst 31(2):101717
81. Faraji A, Rezaei F, Rahnamayiezekavat P, Rashidi M, Soleimani H (2023) Subjective and simulation-based analysis of discomfort glare metrics in office buildings with light shelf systems. Sustainability 15(15):11885
82. Adel A (2022) Future of industry 5.0 in society: human-centric solutions, challenges and prospective research areas. J Cloud Comput 11(1):40
83. Kaur DP, Singh NP, Banerjee B (2023) A review of platforms for simulating embodied agents in 3D virtual environments. Artif Intell Rev 56(4):3711–3753
84. Mourtzis D, Angelopoulos J, Panopoulos N (2022) Operator 5.0: a survey on enabling technologies and a framework for digital manufacturing based on extended reality. J Mach Eng 22:43
85. Díez-González J, Verde P, Ferrero-Guillén R, Álvarez R, Juan-González N, Martínez-Gutiérrez A (2023) Evaluation of the skills' transfer through digital teaching methodologies. In: Computational intelligence in security for information systems conference. Springer, pp 340–349
86. Rezaei F, Faraji A (2024) Performance assessment of light shelves system in office buildings. In: The second international conference on art & architecture, advanced technologies, & construction management (IAAC), pp 1–15
87. Bhattacharya P et al (2023) Towards future internet: the Metaverse perspective for diverse industrial applications. Mathematics 11(4):941
88. Sim JK et al (2024) Designing an educational Metaverse: a case study of NTUniverse. Appl Sci (Switz) 14(6). https://doi.org/10.3390/app14062559
89. Hummel M, van Kooten K (2019) Leveraging NVIDIA Omniverse for in situ visualization. In: High performance computing: ISC high performance 2019 International workshops, Frankfurt, Germany, June 16–20, 2019, revised selected papers 34. Springer, pp 634–642
90. USD documentation: open source release. https://openusd.org/docs/Open-Source-Release.html. Accessed 12 Aug 2024
91. Amaizu GC, Njoku JN, Lee J-M, Kim D-S (2024) Metaverse in advanced manufacturing: background, applications, limitations, open issues & future directions. ICT Express 10:233
92. Digital Twins and NVIDIA Omniverse: the future of infrastructure planning—Kodifly. https://kodifly.com/digital-twins-and-nvidia-omniverse. Accessed 31 Aug 2025
93. Saback V, Popescu C, Blanksvärd T, Täljsten B (2024) Analysis of digital twins in the construction industry: practical applications, purpose, and parallel with other industries. Buildings. https://doi.org/10.3390/buildings14051361
94. Faraji A, Ghasemi E, Homayoon Arya S, Abagheri Mahabadi H. Prediction of effects of Industry 5.0 concepts on construction 5.0. https://www.researchgate.net/publication/385848718

Feature Selection and Multicollinearity-Sensitivity Relationship in Tunnel Boring Machine Performance Assessment

Jitendra Khatti and Swapnil Mishra

Abstract The assessment of tunnel boring machine (TBM) performance parameters is crucial for the successful planning and execution of tunneling projects. This study focuses on predicting the advance rate (AR), penetration rate (PR), and field penetration index (FPI) of shield TBMs using geological and operational parameters. A database was compiled from the literature comprising cutterhead rotation speed (CRS), mean thrust (MF), mean cutterhead torque (MT), upper earth pressure (UEP), torque penetration index (TPI), specific energy (SE), and lower earth pressure (LEP). The variance inflation factor (VIF) method was applied to evaluate the multicollinearity levels of these features, and four combinations were developed accordingly. Performance analysis revealed that empirical models trained with the full feature set (CRS, MF, MT, UEP, TPI, SE, and LEP) achieved the most reliable predictions of AR, PR, and FPI. Furthermore, the feature multicollinearity-sensitivity relationship indicated that the relative multicollinearity of operational and geological parameters had a less significant impact on prediction accuracy and curve fitting.

Keywords Advance rate · Penetration rate · Multicollinearity-sensitivity relationship plot · Feature selection · Curve fitting

J. Khatti (✉)
Department of Civil Engineering, Rajasthan Technical University, Kota, Rajasthan, India

S. Mishra
Department of Mining Engineering, Indian Institute of Technology (Indian School of Mines), Dhanbad, Jharkhand, India
e-mail: swapnil@iitism.ac.in

M. Mirrashid et al. (eds.), *Optimization and Adaptive Control Algorithms*, Emerging Trends in Mechatronics, https://doi.org/10.1007/978-981-95-8387-4_5

1 Introduction

For mechanized tunneling projects, precise performance assessment of the TUNNEL BORING MACHINE (TBM) is essential for scheduling, equipment selection, planning, and cost control [1]. TBMs' versatility in excavating under a range of geological circumstances makes them popular for usage in urban, transportation, and hydropower tunnels. Indicators like the penetration rate (PR), which shows cutterhead penetration each revolution, and the advance rate (AR), which shows tunnel advancement, are the mainstays of performance measurement. Important information on TBM–ground interaction is also provided by the field penetration index (FPI), which connects thrust force to penetration, and boreability, which characterizes the difficulty of excavation [2, 3]. For early TBM performance prediction, empirical models, such as the QTBM classification, the Norwegian University of Science and Technology (NTNU) model, and those developed from testing and field data by the Colorado School of Mines (CSM), are essential.

Yagiz [4] achieved a correlation coefficient of 0.82 with a method that includes disc penetration per revolution (DPW), angle (α), uniaxial compressive strength (UCS), and penetration strength index (PSI). Volumetric joint count (J_v) and UCS have a significant impact on penetration rate (PR), according to Gong and Zhao [5], who also pointed out that PR and the brittle index rise concurrently, while UCS is inversely proportional to PR. They discovered that at angles less than 60°, PR increases as the tunnel axis is oriented to the joint planes. Strong relationships between rock mass characteristics (RMCs), field penetration index (FPI), and fracture circumstances were revealed by Hassanpour et al. [6, 7]. Ma and Luo [8] showed that tunneling costs may be reliably predicted using empirical approaches. With coefficients greater than 0.80, Yagiz et al. [9] created nonlinear equations utilizing UCS, Bi, DPW, and α. Furthermore, Shahriar et al. [10] showed that PR has a high correlation with UCS, rises with UCS and joint spacing (J_s), and falls with volumetric joint count (J_v). The performance of the TBM was measured using geological and operational data, and the following equations were developed [11–17]:

$$\begin{aligned}\mathrm{PR} = {} & 3.425 + 0.03 * \mathrm{RPM} + 0.001 * \mathrm{TF} - 0.096 * \mathrm{WZ} \\ & -0.021 * \mathrm{BTS} - 0.02 * \mathrm{RMR} - 0.003 * \mathrm{UCS} - 0.004 * \mathrm{RQD}\end{aligned} \tag{1}$$

$$\mathrm{FPI} = 0.0066\left(\mathrm{DRI}^2 \ln\left(\mathrm{DRI}\right)\right) + 82.812 * \left(1/\sqrt{\propto^3}\right) + 0.0009 * J_S^2 + 22.348 \tag{2}$$

$$\begin{aligned}\mathrm{PR} = {} & 0.01312 * \propto^{0.013121} - 0.32839 * \mathrm{DPW} \\ & + 0.029086 * \mathrm{Bi} - 0.0035507 * \mathrm{UCS} - 12.661\end{aligned} \tag{3}$$

$$\mathrm{PR} = 5.42 - 0.013 * \mathrm{UCS} - 0.003 * \alpha + 0.066 * \mathrm{RPM} + 0.001 * T - 0.00008 * F \tag{4}$$

$$\mathrm{PR} = 2.602 - 0.012 * \mathrm{UCS} + 0.037 * J_v + 0.009 * \alpha - 0.004 * J_s \tag{5}$$

$$\mathrm{FPI} = 63.267e^{0.847*S_p} \tag{6}$$

$$\mathrm{PR} = 0.067 + 0.485*\alpha^{0.231} - 0.231*\mathrm{DPW} + 0.541*\mathrm{Bi} - 0.192*\mathrm{UCS} \tag{7}$$

$$\mathrm{FPI} = 11.85 + 0.116*\mathrm{RMC} \tag{8}$$

$$\mathrm{BI} = \left(e^{-0.09\sin(\propto+30)} + 0.84e^{-0.05J_v}\right)*3706*\mathrm{UCS}^{0.26}\mathrm{Bi}^{-0.10} \tag{9}$$

$$\mathrm{PR} = -0.219*\mathrm{DPW} + 0.437*\log(\alpha) - 0.003*\mathrm{UCS} + 0.029*\mathrm{PSI} + 1.093 \tag{10}$$

where RPM is the rounds per minute, TF is the thrust force, WZ is the weathering zone, BTS is the Brazilian tensile strength, RMR is the rock mass rating, RQD is the rock quality designation, DRI is the drilling rate index, $\propto$ is the angle between the tunnel axis and the planes of weakness, Bi is the boreability index, *T* is the torque, *F* is the thrust per cutter, and Sp is the spacing. Equations (1–10) have been used to estimate the performance of TBMs, and the correlation coefficients show a range of 0.75–0.90. Geological and operational characteristics served as the main foundation for the development of these equations. It is important to note, though, that a number of crucial factors have not been taken into account when assessing the penetration rate (PR), advance rate (AR), and field penetration index (FPI) of TBMs. These factors include the cutterhead rotation speed (CRS), mean thrust (MF), mean cutterhead torque (MT), upper earth pressure (UEP), torque penetration index (TPI), specific energy (SE), and lower earth pressure (LEP). Additionally, feature multicollinearity and its effects on model performance and possible overfitting in empirical approaches have not been sufficiently covered in prior research. To address these gaps, this study introduces the following key novelties:

- This study develops empirical models that incorporate key geological and operational parameters, i.e., CRS, MF, MT, UEP, TPI, SE, and LEP, to predict the PR, AR, and FPI of shield TBMs.
- The investigation quantifies feature multicollinearity using the variance inflation factor (VIF) method and constructs different feature combinations based on their levels of multicollinearity to examine their influence on the overfitting behavior of empirical models.
- For the first time, this research explores the overfitting tendencies of empirical models by introducing and analyzing a feature multicollinearity-sensitivity relationship plot.

This investigation provides practical insights for tunnel engineers and designers in estimating the AR, PR, and FPI of shield TBMs. By considering various combinations of geological and operational parameters, engineers can more accurately assess the performance of the TBM. For this purpose, a comprehensive database comprising CRS, MF, MT, UEP, TPI, SE, LEP, PR, AR, and FPI from 1197 tunnel stations has been compiled from the literature [18]. After preprocessing, i.e., removing outliers using the z-score method and addressing missing values, 1159 reliable data points have been obtained. The screened database has been subjected to descriptive statistical analysis, distance correlation, multicollinearity assessment, and sensitivity evaluation. Subsequently, the dataset has been randomly partitioned into training (85%, $n = 985$) and testing (15%, $n = 174$) subsets. Different feature combinations have been created based on their levels of multicollinearity and used to develop multiple linear regression (MLR) models. Model performance has been evaluated using root mean square error (RMSE), mean absolute error (MAE), and correlation coefficient (R). The comparative analysis identifies the most influential features for predicting PR, AR, and FPI. Furthermore, model overfitting has been systematically examined through a novel feature multicollinearity-sensitivity relationship plot.

2 Database Insights and Analysis

The Guangzhou–Foshan intercity railway project database [18] is used in this study to create and examine empirical models. The Longdong–Dayuan (Long–Da) section of the 46.5 km Guangzhou South–Zhuliao railway line consists of mining and shield tunnels. A 9.15 m cutterhead and a 9.1 m trailing diameter shield machine were used to dig the about 3.11-km-long shield tunnel. To provide a reaction force for hydraulic jacks, precast concrete lining rings were erected inside the shield. They were made up of six segments and a key piece that measured 8.0 m in inner diameter, 8.8 m in outer diameter, 1.8 m in breadth, and 0.4 m in thickness. Although soil and rock tests were conducted using boreholes, the interpolated data only provided a general description of the tunnel geology by approximating in situ geological conditions. CRS, MF, MT, UEP, TPI, SE, LEP, PR, AR, and FPI parameters for 1197 tunnel stations are included in the database. These parameters have been preprocessed in order to create empirical models. The statistical overview of the preprocessed data points, or 1159, is shown in Table 1. As shown in Fig. 1, the distance correlation approach shows the relationship between characteristics and labels as well as the database's quality. The distance correlation matrix shows how performance labels (PR, AR, and FPI) and features (CRS, MF, MT, UEP, LEP, TPI, and SE) relate to one another. There is a moderate correlation between CRS and FPI ($r = 0.424$), PR ($r = 0.537$), and AR ($r = 0.457$). While MF and MT have strong correlations with FPI (0.720, 0.626), they also have strong correlations with PR (0.507, 0.483) and AR (0.483, 0.473). While UEP and LEP have moderate connections with FPI (0.420, 0.526), they have weak-to-moderate relationships with PR (0.342, 0.396) and AR (0.309, 0.361). TPI and SE, on the other hand, show very strong correlations with FPI (0.906, 0.907), PR (0.823), and AR (0.819, 0.818). In terms of forecasting TBM performance metrics, TPI and SE stand out as the most significant characteristics.

The variance inflation factor ($VIF = 1/(1 - R^2)$), where R^2 is the coefficient of determination, was employed to assess the multicollinearity of TBM features. The results indicate that CRS (VIF = 1.31), MT (VIF = 4.52), and UEP (VIF = 8.15)

Table 1 Statistical summary of the preprocessed database

Variables	Unit	Category	Mean	StDev	Kurtosis	Skewness	Min	Max	CL
CRS	RPM	Feature	1.43	0.17	−0.19	0.72	0.90	1.90	0.01
MF	F/A	Feature	449.91	167.18	−0.67	0.45	66.91	882.00	9.63
MT	T/D^3	Feature	6.58	3.28	−0.21	0.79	1.57	16.71	0.19
UEP	MPa	Feature	0.16	0.08	−0.27	−0.54	0.00	0.33	0.00
LEP	MPa	Feature	0.20	0.08	−0.49	−0.18	0.00	0.40	0.00
TPI	–	Feature	512.92	426.94	0.54	1.03	48.75	1955.56	24.61
SE	kW h/ m^3	Feature	13.73	11.35	0.53	1.03	1.37	52.09	0.65
PR	mm/r	Label	16.60	11.15	−0.43	0.84	2.35	48.46	0.64
AR	mm/ min	Label	22.66	14.07	−0.50	0.81	4.00	63.00	0.81
FPI	–	Label	2889.06	2149.89	−0.27	0.78	327.28	9988.33	123.90

Note: StDev represents the standard deviation, and CL denotes the 95% confidence level

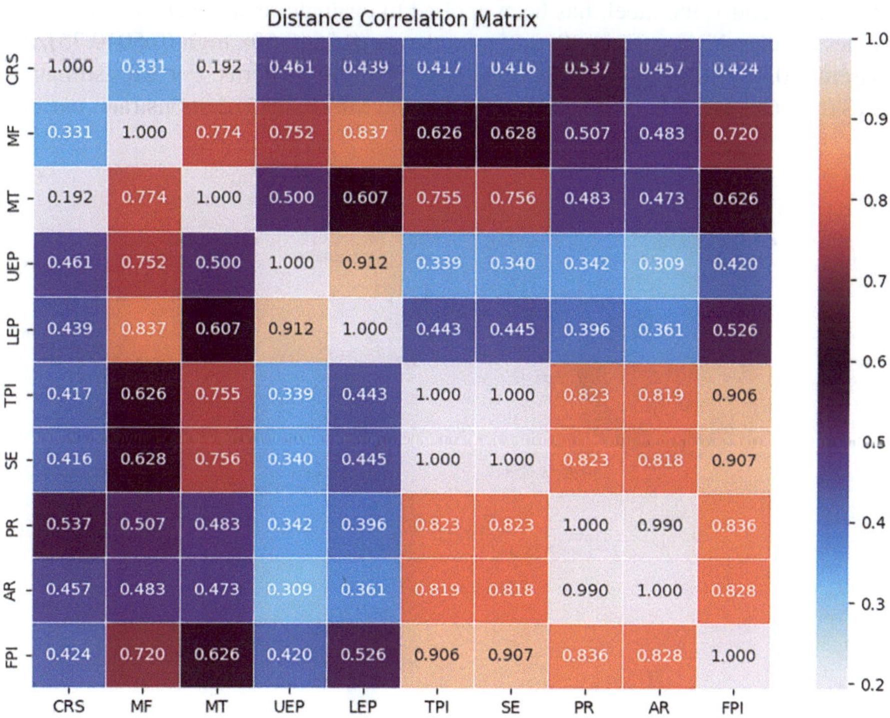

Fig. 1 Illustration of the relationship between variables using the distance correlation method

exhibit weak, considerable, and moderate multicollinearity, respectively [19]. LEP exhibits problematic multicollinearity in estimating PR, AR, and FPI, whereas MF, TPI, and SE display infinite VIF values, indicating severe multicollinearity. Figure 1 further demonstrates strong to very strong interdependencies among MF, TPI, SE, MT, UEP, and LEP. Specifically, MF strongly correlates with MT (0.774), UEP (0.752), and LEP (0.837); TPI correlates strongly with MT (0.755) and moderately with UEP (0.339) and LEP (0.443), while SE shows very strong correlation with MT (0.756) and moderate correlation with UEP (0.340) and LEP (0.445). Collectively, these findings suggest that cutterhead torque has a strong influence on energy-related indices, while earth pressures exert a moderate effect. Thus, a ***relative multicollinearity*** exists between the group of MF, TPI, and SE (highly interdependent operational energy features) and the group of MT, UEP, and LEP (torque–pressure features), highlighting their overlapping explanatory power in predicting TBM performance.

Based on the correlation matrix and multicollinearity analysis, it has been observed that CRS, MF, MT, UEP, TPI, SE, and LEP play significant roles in estimating TBM performance parameters. To further quantify their influence, the cosine amplitude sensitivity method ($\mathrm{CASM} = \sum_{i=1}^{N} x_i \cdot y_i \,/\, \sqrt{\sum_{i=1}^{N} x_i \cdot \sum_{i=1}^{N} y_i}$), where x_i represents the feature and y_i the label, has been applied to evaluate feature sensitivity. CASM values are classified as weak (0–0.25), moderate (0.25–0.50), high (0.50–0.75), and extreme (0.75–1.0). Figure 2 illustrates the sensitivity patterns. For AR estimation, CRS (0.8161), UEP (0.7551), LEP (0.7451), and MF (0.7285) demonstrates extreme

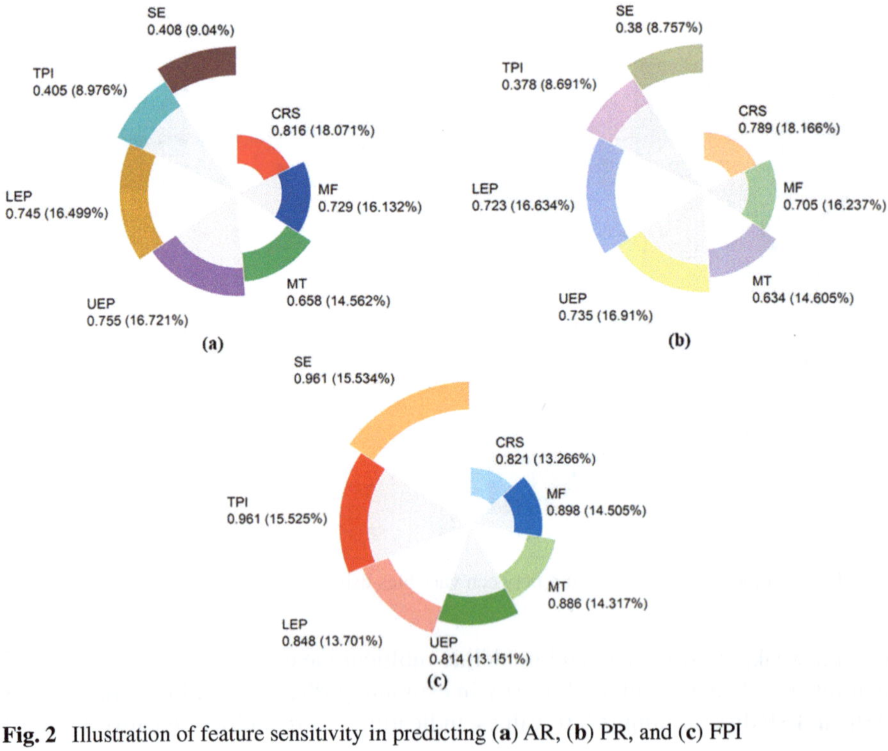

Fig. 2 Illustration of feature sensitivity in predicting (**a**) AR, (**b**) PR, and (**c**) FPI

sensitivity, contributing 18.071%, 16.721%, 16.499%, and 16.132%, respectively (refer to Fig. 2a). Similarly, PR estimation (refer to Fig. 2b) is strongly influenced by CRS, UEP, LEP, and MF. For FPI prediction (refer to Fig. 2c), all features, i.e., SE, TPI, MF, MT, LEP, CRS, and UEP, present extreme sensitivity, with nearly equal contributions (13–15.5%). These results confirm that both energy-related and torque–pressure parameters critically drive TBM performance predictions.

3 Development of Empirical Models

This study employs multilinear regression (MLR) models to estimate the AR, PR, and FPI parameters of the shield TBM. MLR is a statistical technique used to model the relationship between a dependent variable and multiple independent variables. It extends simple linear regression by incorporating additional predictors, enabling more accurate estimation of complex real-world problems. The general equation for multilinear regression is

$$T = C_0 + C_1X_1 + C_2X_2 + C_3X_3 + \ldots + C_nX_n + E \tag{11}$$

where T is the label, C_i are coefficients, X_n are the features, and E is the error. The multicollinearity analysis reveals that (a) CRS has weak, (b) MT has considerable, (c) UEP has moderate, (d) LEP has problematic, and (e) MF, TPI, and SE have relative problematic multicollinearity. Therefore, four feature combinations have been created: (1) Case1: CRS, MT, and UEP; (2) Case2: CRS, MT, UEP, and LEP; (3) Case3: MF, LEP, TPI, and SE; and (4) Caes4: CRS, MF, MT, UEP, LEP, TPI, and SE to analyze the feature multicollinearity impact on performance and overfitting of the empirical models. The following equations (Eqs. (12–15) for AR, Eqs. (16–19) for PR, and Eqs. (20–23) for FPI prediction) were derived during the training phase:

$$\mathrm{AR}_{\mathrm{case1}} = 74.5889 - 29.5349 * \mathrm{CRS} - 22.06291 * \mathrm{MT} + 24.2311 * \mathrm{UEP} \tag{12}$$

$$\begin{aligned}\mathrm{AR}_{\mathrm{case2}} = {} & 73.5158 - 28.1123 * \mathrm{CRS} - 1.82554 * \mathrm{MT} \\ & + 72.0264 * \mathrm{UEP} - 51.2934 * \mathrm{LEP}\end{aligned} \tag{13}$$

$$\begin{aligned}\mathrm{AR}_{\mathrm{case3}} = {} & 32.9626 - 0.00763444 * \mathrm{MF} + 30.3425 * \mathrm{LEP} \\ & - 0.0246588 * \mathrm{TPI} - 0.000656256 * \mathrm{SE}\end{aligned} \tag{14}$$

$$\begin{aligned}\mathrm{AR}_{\mathrm{case4}} = {} & 47.3948 - 10.0013 * \mathrm{CRS} - 0.0264676 * \mathrm{MF} \\ & + 1.81032 * \mathrm{MT} + 54.267 * \mathrm{UEP} - 21.1696 * \mathrm{LEP} \\ & - 0.0286868 * \mathrm{TPI} - 0.000768339 * \mathrm{SE}\end{aligned} \tag{15}$$

$$\mathrm{PR}_{\mathrm{case1}} = 66.9663 - 29.6642 * \mathrm{CRS} - 1.5352 * \mathrm{MT} + 13.1272 * \mathrm{UEP} \tag{16}$$

$$\begin{aligned}\mathrm{PR}_{\mathrm{case2}} = {} & 66.1717 - 28.6108 * \mathrm{CRS} - 1.35936 * \mathrm{MT} \\ & + 48.5169 * \mathrm{UEP} - 37.9799 * \mathrm{LEP}\end{aligned} \tag{17}$$

$$\begin{aligned}\mathrm{PR}_{\mathrm{case3}} = {} & 24.9099 - 0.00544604 * \mathrm{MF} + 21.3758 * \mathrm{LEP} \\ & - 0.0193552 * \mathrm{TPI} - 0.000514958 * \mathrm{SE}\end{aligned} \tag{18}$$

$$\begin{aligned}\mathrm{PR}_{\mathrm{case4}} = {} & 47.4044 - 15.532 * \mathrm{CRS} - 0.0201946 * \mathrm{MF} \\ & + 1.28082 * \mathrm{MT} + 36.0961 * \mathrm{UEP} - 15.2457 * \mathrm{LEP} \\ & - 0.0206417 * \mathrm{TPI} - 0.000553181 * \mathrm{SE}\end{aligned} \tag{19}$$

$$\mathrm{FPI}_{\mathrm{case1}} = -5760.81 + 4035.26 * \mathrm{CRS} + 347.928 * \mathrm{MT} + 3898.36 * \mathrm{UEP} \tag{20}$$

$$\begin{aligned}\mathrm{FPI}_{\mathrm{case2}} = {} & -5568.48 + 3780.28 * \mathrm{CRS} + 305.364 * \mathrm{MT} \\ & - 4668.44 * \mathrm{UEP} + 9193.8 * \mathrm{LEP}\end{aligned} \tag{21}$$

$$\begin{aligned}\mathrm{FPI}_{\mathrm{case3}} = {} & -505.211 + 4.55246 * \mathrm{MF} - 2667.52 * \mathrm{LEP} \\ & + 3.62793 * \mathrm{TPI} + 0.0975042 * \mathrm{SE}\end{aligned} \tag{22}$$

$$\begin{aligned}\mathrm{FPI}_{\mathrm{case4}} = {} & -850.529 + 319.641 * \mathrm{CRS} + 8.13631 * \mathrm{MF} \\ & - 430.994 * \mathrm{MT} - 2426.14 * \mathrm{UEP} + 645.838 * \mathrm{LEP} \\ & + 5.26936 * \mathrm{TPI} + 0.142043 * \mathrm{SE}\end{aligned} \tag{23}$$

4 Results and Discussion

This study develops four empirical models to estimate each AR, PR, and FPI parameter of the shield TBM. The prediction capabilities of each model have been measured and analyzed by implementing $\mathrm{RMSE} = \sqrt{1/n\sum_{i=1}^{n}(A_i - P_i)^2}$, $\mathrm{MAE} = 1/n\sum_{i=1}^{n}|A_i - P_i|$, and $R = \sum_{i=1}^{n}(A_i - A_{\mathrm{mean}})(P_i - P_{\mathrm{mean}}) / \sqrt{\sum_{i=1}^{n}(A_I - A_{\mathrm{mean}})^2 \sum_{i=1}^{n}(P_i - P_{\mathrm{mean}})^2}$ metrics, where A_i is the actual ith value, P_i is the predicted ith value, A_{mean} is the mean of actual values, and P_{mean} is the mean of predicted values. An empirical model is called robust if it achieves RMSE = 0, MAE = 0, and $R = 1$ in both phases. Table 2 presents the training and testing performances of empirical models in assessing AR, PR, and FPI parameters. In the AR prediction, $\mathrm{AR}_{\mathrm{Case4}}$ demonstrates superior accuracy with the lowest RMSE (8.52 training, 8.61 testing) and highest R (0.7980 training, 0.7784 testing), indicating strong reliability. PR models also show progressive improvement, with $\mathrm{PR}_{\mathrm{Case4}}$ achieving the best performance (RMSE = 6.50 training, 6.64 testing; R = 0.8142 training, 0.7964 testing). Similarly, FPI models exhibit

Table 2 Summary of training and testing performances of the empirical models

	RMSE	MAE	R	RMSE	MAE	R
Models	Training performance			Testing performance		
AR_{Case1}	11.24	8.94	0.6067	11.11	8.67	0.5767
AR_{Case2}	11.15	8.86	0.6152	11.11	8.57	0.5761
AR_{Case3}	9.39	7.57	0.7477	9.37	7.62	0.7294
AR_{Case4}	8.52	6.85	0.7980	8.61	6.95	0.7784
PR_{Case1}	8.40	6.64	0.6614	8.38	6.44	0.6401
PR_{Case2}	8.33	6.59	0.6683	8.38	6.34	0.6401
PR_{Case3}	7.51	6.00	0.7419	7.64	6.20	0.7183
PR_{Case4}	6.50	5.18	0.8142	6.64	5.31	0.7964
FPI_{Case1}	1511.00	1116.72	0.7103	1560.73	1206.06	0.6944
FPI_{Case2}	1488.62	1093.68	0.7205	1558.22	1201.17	0.6946
FPI_{Case3}	810.70	544.90	0.9260	802.24	534.53	0.9321
FPI_{Case4}	407.86	281.89	0.9818	428.41	282.02	0.9806

remarkable predictive strength, particularly in Case4, which achieved the lowest RMSE (407.86 for training and 428.41 for testing) and a near-perfect correlation ($R \approx 0.98$). The training equations reveal that model accuracy improved as more relevant features (e.g., CRS, UEP, LEP, SE, TPI) are integrated, enhancing sensitivity to key geotechnical variables. Overall, the results confirm that models with higher feature inclusion (AR_{Case4}, PR_{Case4}, and FPI_{Case4}) yield superior predictive performance and generalization across both training and testing phases. Figure 3 illustrates the comparison of actual and predicted AR, PR, and FPI parameters of shield TBM using Eqs. (15), (19), and (23), respectively.

The test-to-train RMSE ratio was calculated to examine the influence of feature multicollinearity on the curve-fitting performance of the empirical models in predicting the AR, PR, and FPI parameters of shield TBMs. As illustrated in Fig. 4a, the models developed using CRS, MF, MT, UEP, LEP, TPI, and SE (Case4) predicted AR (= 1.0095), PR (= 1.0218), and FPI (= 1.0329), exhibiting slight overfitting yet a close approximation to an ideal fit. This overfitting tendency can be attributed to the multicollinearity introduced by the LEP feature, which appears to influence the models' performance significantly. In contrast, the relative multicollinearity effects of operational energy and torque–pressure features were less significant in Case4, as shown in Fig. 4b. Furthermore, Fig. 4b highlights that the empirical models yield more stable predictions with reduced error and minimal overfitting when CRS, MT, and UEP features exhibit VIF values below 10, thereby ensuring acceptable collinearity levels.

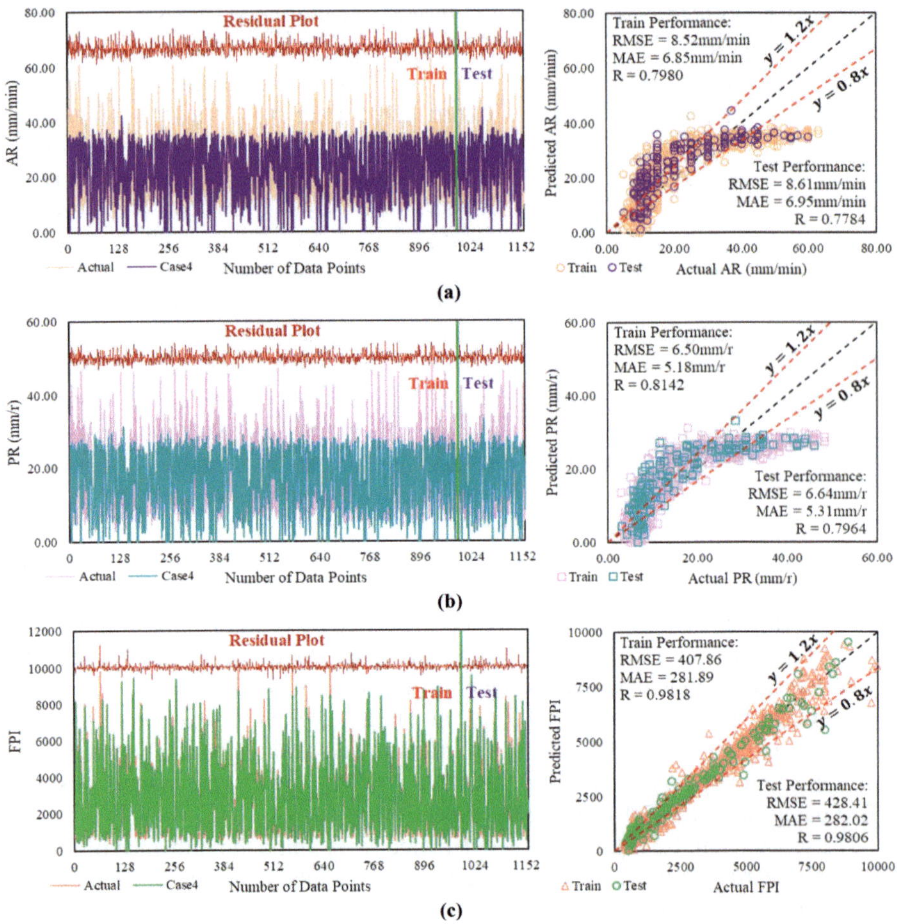

Fig. 3 Illustration of comparison of actual and predicted (**a**) AR, (**b**) PR, and (**c**) FPI

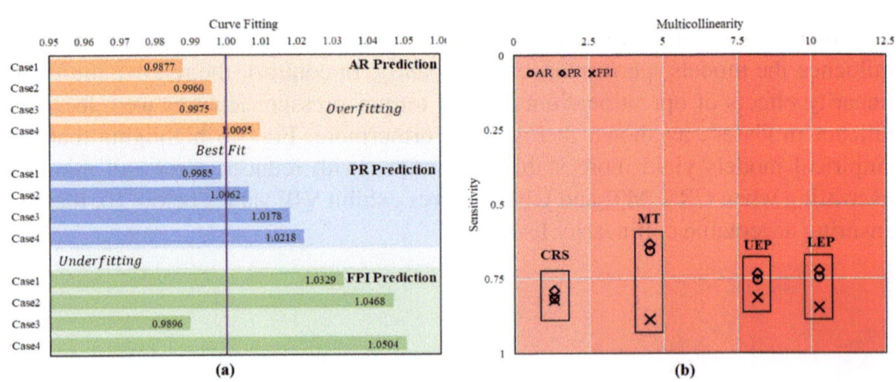

Fig. 4 Illustration of (**a**) curve fitting for each model and (**b**) multicollinearity-sensitivity relationship plot

5 Summary and Conclusion

This study employed empirical models by generating different combinations of CRS, MF, MT, UEP, LEP, TPI, and SE features, selected based on their levels of multicollinearity. A literature-based database was compiled and divided into training and testing sets. Four cases of feature combinations were formulated to predict the AR, PR, and FPI performance parameters of shield TBM. Model performance was evaluated using RMSE, MAE, and *R* metrics. The results demonstrated that AR_{Case4}, PR_{Case4}, and FPI_{Case4} models, which incorporated CRS, MF, MT, UEP, TPI, SE, and LEP features, achieved superior accuracy with minimal residuals in both training and testing phases. Multicollinearity analysis revealed that models developed with only CRS, MT, and UEP tended to underfit due to weak-to-moderate multicollinearity. In contrast, the relative effects of operational energy and torque–pressure features were less significant.

In conclusion, reliable prediction of AR, PR, and FPI parameters can be obtained using the full feature set. However, this study is limited by the number of input variables considered. Future work may incorporate additional geological and operational parameters and extend the methodology to machine learning models for improved predictive capacity. The findings are expected to assist tunnel engineers and designers in more accurately estimating the performance parameters of the shield TBM.

AI Use Disclosure Statement The authors did not use any generative AI or AI-assisted technologies in the creation of this manuscript.

Statements and Declarations

Funding The author declares that no funds, grants, or other support were received during the preparation of this manuscript.

Competing Interests The authors have no relevant financial or non-financial interests to disclose.

Contribution Statement **JK and SM**: Main author, conceptualization, literature review writing and analysis, model development, manuscript preparation, methodological development, statistical analysis, detailing, editing, and overall analysis.

Data Availability The details of the database are provided in the manuscript. The models and codes developed for this research are available from the corresponding author upon reasonable request.

References

1. Ozdemir L (1977) Development of theoretical equations for predicting tunnel boreability, 1970–1979. Mines Theses & Dissertations
2. Howarth DF, Adamson WR, Berndt JR (1986) Correlation of model tunnel boring and drilling machine performances with rock properties. Int J Rock Mech Min Sci Geomech Abstracts 23(2):171–175. https://doi.org/10.1016/0148-9062(86)90344-X

3. Bruland A (1998) Hard rock tunnel boring. Ph.D. Thesis, Norwegian University of Science and Technology, Trondheim
4. Yagiz S (2008) Utilizing rock mass properties for predicting TBM performance in hard rock condition. Tunn Undergr Space Technol 23(3):326–339. https://doi.org/10.1016/j.tust.2007.04.011
5. Zhao Z, Gong Q, Zhang Y, Zhao J (2007) Prediction model of tunnel boring machine performance by ensemble neural networks. Geomech Geoengin 2(2):123–128. https://doi.org/10.1080/17486020701377140
6. Hassanpour J, Rostami J, Khamehchiyan M, Bruland A (2009) Developing new equations for TBM performance prediction in carbonate-argillaceous rocks: a case history of Nowsood water conveyance tunnel. Geomech Geoeng Int J 4(4):287–297. https://doi.org/10.1080/17486020903174303
7. Hassanpour J, Rostami J, Khamehchiyan M, Bruland A, Tavakoli HR (2010) TBM performance analysis in pyroclastic rocks: a case history of Karaj water conveyance tunnel. Rock Mech Rock Eng 43:427–445. https://doi.org/10.1007/s00603-009-0060-2
8. Ma J, Luo X (2009) The time and cost modeling of TBM in tunnelling based on risk evaluation. In: 2009 Third international symposium on intelligent information technology application workshops. IEEE, pp 202–205. https://doi.org/10.1109/IITAW.2009.129
9. Yagiz S, Gokceoglu C, Sezer E, Iplikci S (2009) Application of two nonlinear prediction tools to the estimation of tunnel boring machine performance. Eng Appl Artif Intell 22(4–5):808–814. https://doi.org/10.1016/j.engappai.2009.03.007
10. Shahriar K, Ahangari K, Kamali-Bandpey H (2012) A statistical model for prediction TBM performance using rock mass characteristics in the TBM driven Alborz tunnel project. Res J Appl Sci Eng Technol 4(23):5048–5054
11. Jahed Armaghani D, Faradonbeh RS, Momeni E, Fahimifar A, Tahir MM (2018) Performance prediction of tunnel boring machine through developing a gene expression programming equation. Eng Comput 34:129–141. https://doi.org/10.1007/s00366-017-0526-x
12. Zare Naghadehi M, Ramezanzadeh A (2017) Models for estimation of TBM performance in granitic and mica gneiss hard rocks in a hydropower tunnel. Bull Eng Geol Environ 76:1627–1641. https://doi.org/10.1007/s10064-016-0950-y
13. Minh VT, Katushin D, Antonov M, Veinthal R (2017) Regression models and fuzzy logic prediction of TBM penetration rate. Open Eng 7(1):60–68. https://doi.org/10.1515/eng-2017-0012
14. Hosseini M, Hosseini SJ (2017) Introducing new equation for predicting penetration rate of tunnel boring machine. Amirkabir J Civ Eng 49(2):313–322
15. Yagiz S, Karahan H (2011) Prediction of hard rock TBM penetration rate using particle swarm optimization. Int J Rock Mech Min Sci 48(3):427–433. https://doi.org/10.1016/j.ijrmms.2011.02.013
16. Hassanpour J, Rostami J, Zhao J (2011) A new hard rock TBM performance prediction model for project planning. Tunn Undergr Space Technol 26(5):595–603. https://doi.org/10.1016/j.tust.2011.04.004
17. Gong Q, Zhao J (2009) Development of a rock mass characteristics model for TBM penetration rate prediction. Int J Rock Mech Min Sci 46(1):8–18. https://doi.org/10.1016/j.ijrmms.2008.03.003
18. Yan T (2022) Data on prediction of geological characteristics during shield tunnelling in mixed soil and rock ground. Data Brief 45:108726. https://doi.org/10.1016/j.dib.2022.108726
19. Khatti J, Grover KS (2023) Prediction of compaction parameters for fine-grained soil: critical comparison of the deep learning and standalone models. J Rock Mech Geotech Eng 15(11):3010–3038. https://doi.org/10.1016/j.jrmge.2022.12.034

Improving Performance of Parsimonious Spiking Neural Networks Used for Few-Shot Image Recognition by Embedding the Function of Astrocyte in Its Structure

Malihe Nazari, Fariba Bahrami, Mohammad Javad Yazdanpanah, and Ali Gholamzade Fard Kazazi

Abstract Spiking neural networks are known as a powerful framework, capable of solving tasks that conventional neural networks are capable of solving, while their dynamics are closer to biological neurons. In other words, spiking neural networks imitate the temporal dynamics of biological neurons, although the role of astrocytes has not often been considered in most models. Inspired by the influence of astrocytes and the concept of the tripartite synapse, we constructed a spiking neural network for learning and incorporated astrocytes to provide feedback to neurons for regulating synaptic strength. Two types of neuron-astrocyte connections were implemented. Using a few-shot learning framework for handwritten digit classification, we demonstrated that neuron-astrocyte interactions improved both classification accuracy and convergence speed of neurons to labels, even with limited training data. Since we employed a few-shot learning approach, we adopted a ten-way ten-shot setup with five query samples per class. Additionally, the network consisted of 80 excitatory and 80 inhibitory neurons. Using this configuration, we achieved an accuracy of 71.95% for the neuronal network, 72.3% for the first configuration where each astrocyte was placed at the synapse between an excitatory and inhibitory neuron, and 72.95% for the second configuration where each astrocyte was placed between two pairs of excitatory and inhibitory neurons. These results highlight the potential of biologically inspired mechanisms to enhance the efficiency and generalization of neuromorphic systems.

M. Nazari (✉) · F. Bahrami · M. J. Yazdanpanah
School of Electrical and Computer Engineering, College of Engineering, University of Tehran, Tehran, Iran
e-mail: nazari.malihe@ut.ac.ir; fbahrami@ut.ac.ir; yazdan@ut.ac.ir

A. G. F. Kazazi
Department of Electrical Engineering, Amirkabir University of Technology (Tehran Polytechnic), Tehran, Iran
e-mail: aghk@aut.ac.ir

M. Mirrashid et al. (eds.), *Optimization and Adaptive Control Algorithms*, Emerging Trends in Mechatronics, https://doi.org/10.1007/978-981-95-8387-4_6

Keywords Spiking neural network · Tripartite synapse · Neuron-astrocyte interaction · Unsupervised learning · STDP · Few-shot learning · Classification Pattern recognition

1 Introduction

The behavior of the nervous system remained a mystery for many years. In other words, understanding the complexities of the nervous system has been one of the major challenges in neuroscience. Furthermore, the details of how the brain performs tasks such as classification and pattern recognition are still unknown. In the classical view of the nervous system, the mistaken belief had been accepted for many years that only neurons are involved in complex cognitive processes, while glial cells only provide an ideal environment for neurons. However, nowadays we know that glial cells respond to neuronal activation by changing their internal Ca^{2+} concentration, which causes the release of their transmitters; as a result, they can regulate neuronal activation by changing synaptic strength. Hence, the tripartite synapse concept has been introduced, which includes the presynaptic neuron, postsynaptic neuron, and surrounding astrocytes (the most common glial cells in the central nervous system) [1–4]. Astrocytes play an essential role in synapse formation and in improving neuronal synchrony by regulating stochastic spiking activity. Furthermore, studies beyond the tripartite synapse have shown that a single astrocyte can surround and influence many synapses from different pre- and postsynaptic neurons [5–7].

Computational modeling is one of the most important parts of neuroscience [8]. Neuronal models are available, and using them has become common. However, this is not true for glial cells. Although it has been proven that they play an important role in neural network activity, most research has not considered them. This means that the significant role of glial cells has been ignored in neuronal physiology. Perhaps this is because there is no standard and unique model for these kinds of glial cells yet.

Spiking neural networks (SNNs), known as the third generation of neural networks, have been receiving increasing attention and are emerging as strong competitors to the previous generations [9–11]. These networks have been introduced as a promising neuromorphic computing framework. They process information only when a spike occurs, which makes them energy-efficient and enhances their performance in processing temporal data, as they operate in an event-driven manner [12]. Their low energy consumption also makes them wellsuited for energy-efficient hardware implementations [13]. Furthermore, owing to their biologically inspired characteristics, SNNs can effectively learn from limited data and generalize well [14].

However, as noted earlier, the role of astrocytes has not been considered in most existing networks, despite their key role in modulating neuronal communication and synaptic plasticity in the brain. In addition, biological systems demonstrate remarkable efficiency, often forming robust representations from only one or a few experiences [15]. Considering this point, we adopted a few-shot learning method. The main objective of this work was to demonstrate the role of astrocytes in neural networks; therefore, we constructed a neural network model for learning. Our focus is on investigating whether astrocytes can enhance the sample efficiency and generalization ability of SNNs under limited training data. We showed that the network's performance is enhanced because astrocytes provide local feedback at the tripartite synapse, while synaptic events propagate to other regions of the network through astrocyte connections.

1.1 Neuron-Astrocyte Interaction Background

The nervous system contains two main kinds of cells: neurons and glial cells. Neurons are the main units of the nervous system. Like any other cell in the body, each neuron has three main parts: nucleus, cytoplasm, and membrane. Each neuron contains a large central part called the soma and some cytoplasmic appendages called axons and dendrites. Neurons are not able to reproduce; as a result, when one of them dies, it has no replacement. A neuron is an electrically excitable cell that processes and exchanges information through both electrical and chemical signals. These signals move through a communication pathway between specialized cells called synapses [16].

The concept of an action potential, or spike, exists in biological neurons. In other words, an action potential is a transient change in a neuron's membrane potential, in such a way that the potential in the membrane rises rapidly because of sodium ion influx and then falls. Neurons' response follows the rule of "all or none." Therefore, a neuron might respond with a firing or spike in response to one or several inputs. When the membrane potential reaches the threshold, the neuron fires, producing a signal and sending it to other neurons. In this way, their potentials increase or decrease; in other words, this action is called a spike of a neuron [17]. The difference between two consecutive spikes is called the inter-spike interval (ISI) [18]. Increasing or decreasing the ISI determines different patterns of neuronal activity. All in all, there are 20 different patterns of neuronal activity, one of which plays a special role in the brain.

As mentioned, the central nervous system contains neurons and glial cells. Neurons, with excitable membranes, are able to produce action potentials; as a result, they are directly involved in transferring nerve signals. On the other hand, glial cells play significant roles, some of which have been discovered in recent years. One of their important roles is to participate in the regulation of synaptic activity. Because of their pumping mechanisms and intracellular enzymes, glial cells are able to transfer key neurotransmitters, such as glutamate, at synapses. Thus, they help in completing the synapse's function. Astrocytes are the most abundant glial cells in the central nervous system. They are star-shaped cells with crucial roles in CNS function [19]. They are interconnected like neurons; however, unlike neurons that communicate through chemical synapses, astrocytes are connected through gap junctions. They also communicate information by releasing ATP. Moreover, they also have neurotransmitter receptors. One of the most important excitatory neurotransmitters is called glutamate. Glutamate is released from the presynaptic neuron into the synaptic cleft, where it activates receptors on the astrocyte. This activation produces the secondary messenger IP_3, which then releases calcium from the endoplasmic reticulum into the cell's cytoplasm. The astrocyte releases glutamate into the synaptic cleft in response to an increase in intracellular calcium ions. Depending on the type of presynaptic receptor, it can release more glutamate, and it can have inhibitory or excitatory effects on the postsynaptic neuron [20, 21]. Neuronal activity also releases ATP, which can affect surrounding astrocytes. However, this effect has two main differences from the previous one: first, it does not depend on calcium release; second, it can affect other astrocytes that are not connected through gap junctions. ATP release can directly excite neurons through the activation of P2X receptors or it can be converted to adenosine, which activates A1 or A2 receptors, leading to an increase or decrease in synaptic activity. Synaptic regulation by glial cells occurs not only in the CNS but also in the PNS. For instance, the release of ACh and ATP from the presynaptic terminal excites Schwann cells in the nerve and muscle junction. This release is independent of the calcium neurotransmitter. The released neurotransmitter on the neuron acts as a negative feedback and decreases the release of ACH. Besides, astrocytes can control synaptic activity by maintaining the balance of extracellular ions (K^+ and H^+), because an excessive increase in K^+ ions causes depolarization of the synaptic terminal [22]. Thus, a synapse contains three parts: presynaptic neuron, postsynaptic neurons, and glial cells. The connection between neurons and glial cells depends not only on synapses but also on the impulse activity of neurons during embryonic and postnatal stages. This impulse activity could affect the growth of myelinating glial cells. In other words, this activity occurs through the release of nonsynaptic neurotransmitters, growth factors, or special signaling molecules. Increasing intracellular calcium in Schwann cells could change the replication pathways of genes that are involved in reproduction and modification of Schwann cells [23].

One of the most important properties of the vertebrate nervous system is that a large number of neurons are connected to each other; through these connections, they can influence and be influenced by others. A synapse is the special area where one neuron connects to another neuron. Normally, one neuron connects to a few thousand other neurons through synapses and also receives synapses from about the same number [24]. There are several reasons for the importance of synapses. Biologically, the structure of synapses is highly organized, and their function is very stable. Computationally, synapses play one of the most important roles in transferring and processing information. They provide the ability to learn, memory, and adaptability for the brain; this is because they are the basic units responsible for the storage and transmission of information. Hence, synapses have the most crucial role in neuronal communication; synaptic dysfunction causes many brain and mental disorders. Clinically, synapses are the main drug targets for treatment and medical care. Structurally, each synapse consists of a presynaptic element, which usually contains an axon and sometimes a dendrite, and a postsynaptic element, which usually contains a part of a dendrite and sometimes an axon, the soma of the cell, or the axonal initial segment. When an astrocyte is present in a synapse, that synapse is called a tripartite synapse. Synapses are divided into two categories: electrical and chemical. Electrical synapses are suitable for transferring rapid signals, and the presynaptic and postsynaptic terminals are very close together in them. The current caused by an action potential enters directly into the postsynaptic neuron through gap junctions, which act like bridges between the two cells. In fact, they connect the cytoplasm of the presynaptic and postsynaptic cells. Chemical synapses can have inhibitory or excitatory effects on the potential of the postsynaptic terminal and can even strengthen synaptic transmission, so that a small presynaptic terminal has a considerable effect on the postsynaptic terminal. Because of these reasons, most synapses in the brain are of the chemical type. In chemical synapses, the cells are separated by a cleft, and the two cells are not connected through bridges or channels. Instead, the action potential in the presynaptic terminal affects the release of chemical transmitters called neurotransmitters. Chemical transmitters spread in the synaptic cleft and then connect to receptor molecules called neuroreceptors on the postsynaptic terminal; in this way, they regulate the opening and closing of ion channels on the membrane of the postsynaptic neuron. This synaptic exchange can generate and then propagate an action potential in the postsynaptic neuron. Neuroreceptors are divided into two categories based on how they control ion channels in the postsynaptic neuron. First, ionotropic receptors are ion channels that can be activated by connecting to the corresponding neurotransmitter. Second, metabotropic receptors affect ion channels indirectly by activating a chain reaction of second messengers [25]. Both types of receptors can produce inhibitory or excitatory effects. The type of this effect depends not only on the neurotransmitter but also on the type of neuroreceptor for the neurotransmitter. Each neurotransmitter can affect different receptors and produce different effects. Considering the diversity of neuroreceptors, a limited number of neurotransmitters can produce numerous synaptic effects [26].

A few years ago, scientists discovered that astrocytes can sense neuronal activity. In other words, when neurotransmitters pass through the synaptic cleft and reach the surrounding elements, such as astrocytic terminals, they can influence astrocyte activity. This hypothesis was proven by special imaging techniques, since astrocyte calcium concentration changed during neuronal activity. The more glutamate concentration increases in the synaptic cleft area, the more calcium concentration rises, and then calcium oscillates in the astrocyte [27]. So, it has been discovered that astrocytes sense neuronal activity and, in response, release gliotransmitters to regulate the surrounding synapses [28]. In addition, an increase in calcium concentration in one astrocyte leads to an increase in others; as a result, it has been discovered that astrocytes interact with each other. When an action potential arrives at the presynaptic terminal, some of the vesicles release the neurotransmitter glutamate into the active area. Glutamate propagates into the synaptic cleft, and part of it reaches the AMPA, kainate, and NMDA receptors on the postsynaptic terminal, making an excitatory postsynaptic potential (EPSP). If the presynaptic activity is high enough, and a minimal amount of glutamate is released into the synaptic cleft, it is likely to propagate glutamate outside of the synapse and connect to the mGluR I receptor on the astrocyte. These receptors will be connected to phospholipase C, leading to the production of IP_3. These molecules will be connected to IP3Rs on the endoplasmic network after increasing IP_3, which leads to the release of calcium ions. When the calcium concentration reaches the threshold, a type of gliotransmitter (glutamate) releases. Glutamate propagates into the extrasynaptic space and connects to the corresponding receptors on pre- and postsynaptic terminals. For instance, glutamate acts on mGluR I, mGluR II, and NMDA receptors in the presynaptic terminal, and the activation of each of these receptors affects synaptic transmission differently. In the presynaptic terminal, the metabotropic receptors are mainly of the mGluR II type, which reduces the likelihood of glutamate release by inhibiting calcium channels. Besides, they are more likely to release glutamate due to the increase in calcium caused by the activation of NMDA receptors. However, some mechanisms dominate over others depending on the activity of the presynaptic neuron and the timescale. For example, NMDA receptors increase the likelihood of the release and dominate over the mGluR I effect at frequencies between 5 and 10 Hz and on a 100 ms timescale. However, after this period, mGluR I receptors dominate over NMDA receptors at all frequencies [29, 30].

Eventually, the calcium wave is the mechanism of astrocytic communication within the astrocytic network. IP_3 molecules propagate to the surrounding astrocytes through gap junctions, leading to an increase in calcium concentration and activating them due to the rise in both calcium concentration and IP_3 level. Thus, the activity of an astrocyte, which is a sign of the activity of a neuron in a region, can be transferred to a distant region through the emission of a calcium wave. This mechanism is much slower than synaptic transmission, with a timescale between a few seconds and minutes [31].

Overall, the diversity of neurons and their various functions in the central nervous system have been well established—for instance, the differences between pyramidal neurons and interneurons. However, astrocyte identity and morphology have not been properly defined. Currently, we know that astrocytes vary in shape, size, and orientation in a neural circuit and even in different regions of the brain. In addition, the perisynaptic astrocytic process (PAP) could affect synaptic strength and plasticity in a single neuron. Furthermore, research beyond the concept of tripartite synapses has shown that a single astrocyte can affect numerous synapses from several pre- and postsynaptic neurons. Astrocytes show plasticity in response to the synaptic activity, which helps to modify surrounding synapses. So it is important to know astrocytic plasticity to understand learning, memory, and other brain functions, because the complexity of astrocytes in the human brain is partly the reason for higher human intelligence [5].

2 Artificial Neuron-Astrocyte Network

A spiking neural network was designed for this study, in which the structure of connections between neurons is similar to the SNN model presented by Diehl et al. [32]. Our proposed network was trained on the MNIST dataset using an unsupervised learning rule called spike-timing-dependent plasticity (STDP). Afterward, we added astrocytes and tested two types of connections between astrocytes and neurons. In other words, we created tripartite synapses, which consist of a presynaptic neuron, a postsynaptic neuron, and astrocytes.

2.1 Structure of the SNN Model

Structurally, the spiking neural network includes two layers: the first layer is the input layer, and the second layer consists of two sublayers. The first sublayer contains excitatory neurons, and the second sublayer contains inhibitory neurons. For the first layer, we generated Poisson spike trains from the MNIST dataset, where each image is 28 × 28 pixels. We used rate coding to generate a Poisson spike train from each pixel. This is because rate coding is the most common encoding technique in spiking neural network models. In this method, each pixel is mapped to a firing rate so that the pixel intensity is converted into a Poisson spike train. The connection from the first layer to the second layer is all-to-all, meaning each pixel is converted into a spike train, which is then connected to every excitatory neuron. In the second layer, the two sublayers contain an equal number of neurons. Each excitatory neuron is connected to a single inhibitory neuron with the same index, while each inhibitory neuron is connected to all excitatory neurons except its counterpart with the same index. This arrangement results in lateral inhibition. These specific connections enable the network to operate as a competitive network, putting excitatory neurons in constant competition.

The neuronal model for excitatory and inhibitory neurons is based on the leaky integrate-and-fire (LIF) model [33], in which the membrane potential, v, is calculated as follows:

$$\tau_m \frac{dV_k}{dt} = E_{Rest} - V_k(t) + I_{Ak}(t) - I_{Gk}(t) \tag{1}$$

In this equation, τ_m, E_{Rest}, I_A, and I_G are the time constant, resting membrane potential, excitatory synaptic current, and inhibitory synaptic current, respectively. When the membrane potential crosses the threshold potential—in other words, when the neuron fires—it is subsequently reset to the reset potential:

$$\tau_{dA} \frac{dI_{Ak}}{dt} = -I_{Ak} + x_{Ak} \tag{2}$$

$$\tau_{rA}\frac{dx_{Ak}}{dt}=-x_{Ak}+\tau_m\left(J_{k-exc}\sum_{exc}\delta\left(t-t_{k-exc}-\tau_L\right)+J_{k-ext}\sum_{ext}\delta\left(t-t_{k-ext}-\tau_L\right)\right) \quad (3)$$

$$\tau_{dG}\frac{dI_{Gk}}{dt}=-I_{Gk}+x_{Gk} \quad (4)$$

$$\tau_{rG}\frac{dx_{Gk}}{dt}=-x_{Gk}+\tau_m\left(J_{k-inh}\sum_{inh}\delta\left(t-t_{k-inh}-\tau_L\right)\right) \quad (5)$$

In these equations, t_{k-ext}, t_{k-inh}, and t_{k-ext} represent the spike times received from excitatory neurons, inhibitory neurons, and external inputs, respectively. $\tau_{dA}(\tau_{dG})$ and $\tau_{rA}(\tau_{rG})$ denote the time constants of excitatory and inhibitory synaptic currents, respectively. In our network, τ_L is considered zero. J_{k-exc}, J_{k-inh}, and J_{k-ext} represent the synaptic weights from excitatory neurons, inhibitory neurons, and external inputs to the kth neuron.

It is worth noting that the threshold voltage of excitatory neurons is calculated as $v_{thresh}+\theta$. After each spike, θ is increased; otherwise, it decays exponentially.

2.2 Learning Rule

A combination of spike-timing-dependent plasticity (STDP) and the effects of excitatory and inhibitory synaptic currents was employed as the learning method. As is well known, implementation of the classical STDP method is inefficient, since it requires summing over all pairs of spikes, which is practically impossible and physiologically unrealistic, as neurons cannot remember all previous spikes [34]. Therefore, we adopted an online implementation of STDP [35], the equations for which are given below:

$$\tau_+\frac{dx_j}{dt}=-x_j+a_+\left(x_j\right)\sum_f\delta\left(t-t_j^f\right) \quad (6)$$

When there is an incoming presynaptic spike, the trace $x_j(t)$ is updated by an amount $a_+(x_j)$; otherwise, it decays exponentially.

When there is an incoming postsynaptic spike, a trace y is updated by an amount $a_-(y)$; otherwise, it decays exponentially:

$$\tau_-\frac{dy}{dt}=-y+a_-\left(y\right)\sum_n\delta\left(t-t^n\right) \quad (7)$$

According to the two equations above, the weight change upon the arrival of a post-synaptic spike depends on the trace left by the presynaptic neuron. Similarly, the synaptic weight upon the arrival of a presynaptic spike depends on the trace left by the postsynaptic neuron (Table 1):

$$\frac{dw_j}{dt} = A_+(w_j)x(t)\sum_n \delta(t-t^n) - A_-(w_j)y(t)\sum_f \delta(t-t_j^f) \tag{8}$$

2.3 Modeling of Astrocytes

Astrocytes constitute the largest population of glial cells within the central nervous system that are interconnected through the synaptic cleft, unlike neurons, which connect to each other via chemical synapses. Evidence shows a connection between neurons and astrocytes. In addition, studies indicate that astrocytes significantly contribute to nervous system physiology. These findings have resulted in the development of a new framework called tripartite synapses, which consist of a presynaptic neuron, a postsynaptic neuron, and the surrounding astrocytes [1]. Each astrocyte is located at the synapse between excitatory and inhibitory neurons. Although astrocytes are unable to generate action potentials, they can regulate neuronal activity by providing feedback. The astrocyte model is based on the Li–Rinzel formulation [36], with equations given below:

$$\frac{d\left[Ca^{2+}\right]}{dt} = J_{Chan} - J_{pump} + J_{leak} \tag{9}$$

$$\frac{d[Ip_3]}{dt} = \frac{IP_3^* - Ip_3}{\tau_{ip3}} + z \tag{10}$$

$$\frac{dq}{dt} = \left(a_2 d_2 \frac{[Ip_3]+d_1}{[Ip_3]+d_3}\right)(1-q) - a_2\left[Ca^{2+}\right]q \tag{11}$$

Here, $[Ca^{2+}]$, IP_3^*, J_{Chan}, J_{pump}, and J_{leak} represent the cytosolic calcium concentration, equilibrium concentration of $[Ip_3]$, calcium flux from the ER to the cytosol via $[Ip_3]$ receptors, pump flux from the cytoplasm to the ER through an ATP-dependent pump, and leak flux from the ER to the cytosol, respectively:

Table 1 Neuron and STDP parameter values

Parameter	Value	Parameter	Value
v_{Reset}	0	A_+	0.001
E_{Rest}	$5e^{-3}$	A_-	0.0001
v_{thresh}	$20e^{-3}$	θ	$0.05e^{-3}$
J_{min}	0	J_{max}	0.25

$$J_{Chan} = c_1 v_1 p_\infty^3 n_\infty^3 q^3 \left(\left[Ca^{2+}\right]_{ER} - \left[Ca^{2+}\right]\right) \tag{12}$$

$$p_\infty = \frac{[Ip_3]}{[Ip_3] + d_1} \tag{13}$$

$$n_\infty = \frac{\left[Ca^{2+}\right]}{\left[Ca^{2+}\right] + d_5} \tag{14}$$

$$\left[Ca^{2+}\right]_{ER} = \frac{c_0 - \left[Ca^{2+}\right]}{c_1} \tag{15}$$

Here, $[Ca^{2+}]_{ER}$ represents the calcium concentration in the ER. In addition, c_0, c_1, d_1 , and d_5 are constant parameters:

$$J_{pump} = \frac{v_3\left[Ca^{2+}\right]^2}{\left[Ca^{2+}\right] + k_3^2} \tag{16}$$

$$J_{leak} = c_1 v_1 \left(\left[Ca^{2+}\right]_{ER} - \left[Ca^{2+}\right]\right) \tag{17}$$

In the above equations, c_1, k_3, v_1, and v_3 are constant parameters. Neurotransmitter accumulation from the presynaptic neuron to the postsynaptic neuron is calculated as follows:

$$z = r_{ip3}\left([T]_1 + [T]_2\right) \tag{18}$$

$$[T] = \frac{1}{1 + \exp\left(-\frac{v(t) - 0.2}{0.02}\right)} \tag{19}$$

Here, z, $[T]_1$, and $[T]_2$ represent the astrocyte input, and the neurotransmitter accumulation of presynaptic and postsynaptic neurons, respectively. When the calcium concentration of an astrocyte crosses the threshold, it releases glutamate onto neurons and subsequently enters a refractory period. Astrocytes send calcium feedback to neurons:

$$i^{astro} = \lambda\left[Ca^{2+}\right] = \begin{cases} \lambda_1\left[Ca^{2+}\right] & \text{for} \quad \text{same index} \\ \lambda_2\left[Ca^{2+}\right] & \text{o.w.} \end{cases} \tag{20}$$

Table 2 Astrocyte parameter values

Parameter	Value	Parameter	Value
r_{ip3}	2	d_2	1.049
τ_{ip3}	0.55	d_3	0.9434
IP_3^*	0.16	d_5	0.08234
a_2	2	k_3	0.1
c_0	2	v_1	6
c_1	0.185	v_2	0.03
d_1	0.13	v_3	0.9

Astrocytes provide $\lambda_1[Ca^{2+}]$ feedback to the excitatory neuron with the same index and $\lambda_2[Ca^{2+}]$ feedback to the other excitatory neurons. Therefore, the equation for the excitatory neuron with the same index is given as follows:

$$\tau_{rA}\frac{dx_{Ak}}{dt} = -x_{Ak} + \tau_m\left(J_{k-exc}\sum_{exc}\delta\left(t - t_{k-exc}\right) + J_{k-ext}\sum_{ext}\delta\left(t - t_{k-ext}\right) + \lambda_1\left[Ca^{2+}\right]\right) \tag{21}$$

The equation for the other excitatory neurons is given as follows (Table 2):

$$\tau_{rG}\frac{dx_{Gk}}{dt} = -x_{Gk} + \tau_m\left(J_{k-inh}\sum_{inh}\delta\left(t - t_{k-inh} - \tau_L\right) + \lambda_2\left[Ca^{2+}\right]\right) \tag{22}$$

2.4 *Classification Method*

Few-shot learning is used to evaluate the network's ability to recognize new classes from a few training examples. This approach follows an *N*-way *K*-shot setup, in which the model is trained on a support set containing *K* labeled samples for each of *N* classes and then tested on a query set (unseen data) from the same classes. To perform classification, we used the correlation coefficient (Pearson's *r*) as a similarity metric between two spike response patterns. With two spike count vectors, the highest value indicates strong similarity. Given two vectors *x* and *y*, the correlation coefficient is defined as

$$r(x,y) = \frac{\sum_i (x_i - \overline{x})(y_i - \overline{y})}{\sqrt{\sum_i (x_i - \overline{x})^2}\sqrt{\sum_i (y_i - \overline{y})^2}} \tag{23}$$

where $\overline{x}$ and $\overline{y}$ are the mean values of the vectors. Each query was assigned to the class of the support sample with the highest correlation.

3 Applications

To investigate the role of astrocytes, we constructed a network for learning consisting of 80 excitatory and 80 inhibitory neurons and applied it to handwritten digit classification. The MNIST dataset, which contains handwritten digit images, was used. It is commonly employed for training networks in image processing tasks. Two configurations for incorporating astrocytes were considered. In the first configuration, one astrocyte was placed at the synapse between each excitatory and inhibitory neuron, resulting in 80 astrocytes in this configuration (Fig. 1). In the second configuration, a single astrocyte was placed between pairs of excitatory and inhibitory neurons [37], resulting in 40 astrocytes in this configuration (see Fig. 2).

We first ran the network without astrocytes, and subsequently with the two configurations including astrocytes. In the few-shot learning experiments, we adopted a ten-way ten-shot setup with five query samples per class. Each training epoch consisted of 40 tasks, and evaluation was performed on 40 unseen test tasks. By using this method, fewer data are required, which reduces the network simulation time. Consequently, the complexity and computational cost are lower. As a result,

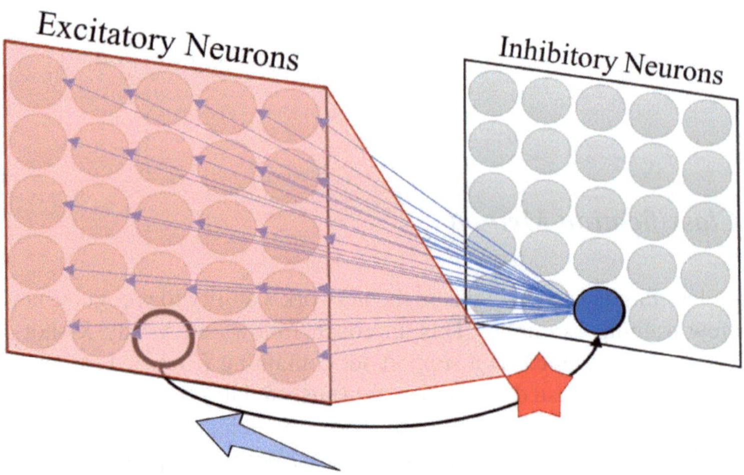

Fig. 1 In this figure, a 5 × 5 neuronal network is shown for simplicity of illustration, whereas the actual network consists of 10 × 8 neurons. The connection from excitatory to inhibitory neurons is one-to-one; in other words, each excitatory neuron connects to the inhibitory neuron with a matching index. Each inhibitory neuron connects to all excitatory neurons except the one with a matching index. Blue arrows represent the connections from inhibitory to excitatory neurons. Given these types of connections, each spike in an excitatory neuron triggers a spike in the inhibitory neuron with the same index. In addition, each inhibitory neuron inhibits all excitatory neurons except the one with a matching index. This action results in lateral inhibition and maintains constant competition. Each astrocyte is located at the synapse between an excitatory and an inhibitory neuron. In the red-shaded area, each astrocyte sends inhibitory feedback to all excitatory neurons except the one with a matching index while providing excitatory feedback to the excitatory neuron with a matching index

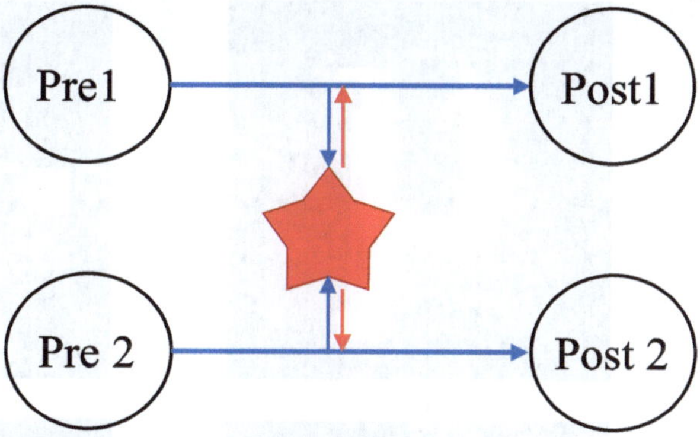

Fig. 2 In this configuration, Pre1 and Pre2 represent two presynaptic neurons, while Post1 and Post2 represent two postsynaptic neurons. Each astrocyte is placed between pairs of excitatory and inhibitory neurons

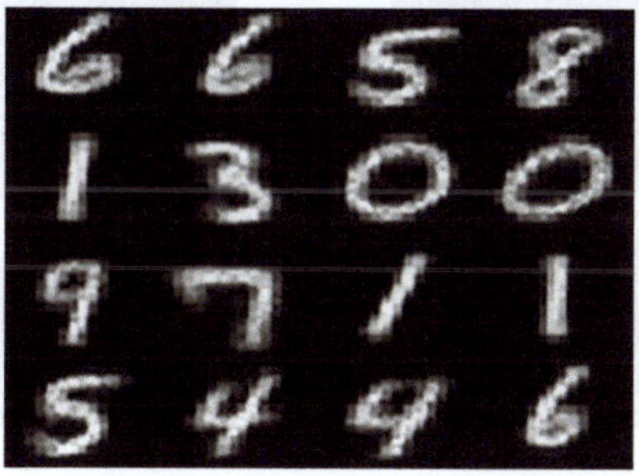

Fig. 3 A part of the receptive field of the network when using one astrocyte between two pairs of excitatory and inhibitory neurons. As illustrated in the figure, pairs of neurons connected through a common astrocyte converge to the same label, indeed with some differences

implementing the network for classification tasks, particularly under conditions with limited data, becomes more feasible. In addition, this approach is more suitable for hardware implementation.

In the spiking neural network (SNN) without astrocytes, an accuracy of 71.95% was obtained. In the network with one astrocyte at each synapse (SNAN), the accuracy increased to 72.3%, and in the network with one astrocyte between two pairs of excitatory and inhibitory neurons, the accuracy reached 72.95%. With the inclusion of astrocytes, the network accuracy increased. This improvement can be attributed to the astrocytes propagating synaptic events to other regions of the network. An important observation is when using the second astrocyte configuration: since each astrocyte monitors two presynaptic and two postsynaptic neurons, it provides excitatory feedback to both excitatory neurons. This mechanism tends to form two equilibrium points for a given label; however, the difference lies in the level of excitation each neuron receives. Because the values of λ differ, one neuron converges directly to the digit it observed, while the other becomes sensitive to the label of that digit, having received only a fraction of the excitatory feedback (Fig. 3).

The confusion matrices are presented in Figs. 5, 6, and 7. Based on the reported accuracies and the confusion matrices, it can be observed that the addition of

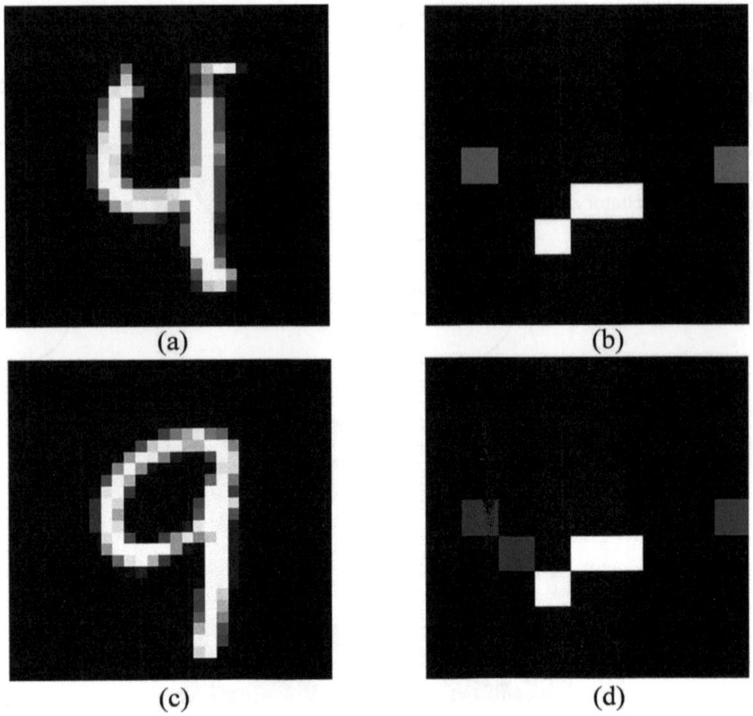

Fig. 4 Example of a digit 4 misclassified as 9 (**a** and **c**). (**b**) The feature map of neurons corresponding to (**a**). (**d**) The feature map of neurons corresponding to (**c**)

astrocytes increases the rate of correct classifications. Additionally, with the mechanism using one astrocyte between two pairs of excitatory and inhibitory neurons, the network's convergence speed is increased. In other words, when the network sees a digit, two neurons become sensitive to that label, and the number of misclassifications is also reduced. The highest number of misclassifications occurs for the digits 4 and 9, followed by 3 and 5, and 8 and 5, which are often misclassified as each other. The reason is almost clear: these digits look similar to each other when handwritten. To illustrate this issue, refer to Fig. 4, which shows a digit 4 that was misclassified as 9. The corresponding feature maps of neurons for both digits are almost similar. In other words, a feature map is the sum of spikes generated by neurons. We observed that the performance of some neurons was highly similar, which led to misclassification.

Two other significant points are that the number of neurons used in our network is 80, while in other spiking neural networks, several thousand neurons are typically

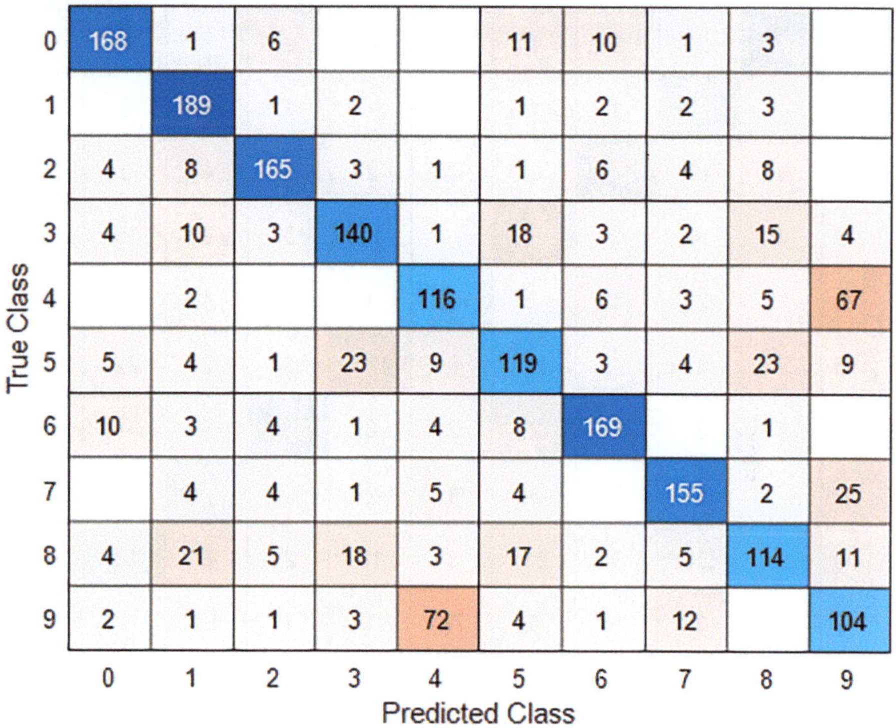

Fig. 5 Confusion matrix of the network without astrocytes

employed. For example, Diehl et al. [32] used 6400 excitatory neurons and inhibitory neurons. Moreover, our network is relatively small, and only a small fraction of the dataset was used. If we increase the number of neurons, the accuracy would likely improve. To demonstrate this, in our SNN, we first used 40 neurons and achieved 68.5% accuracy, and then with 80 neurons, the accuracy increased to 71.95%. As a result, increasing both the number of neurons and the amount of data would further enhance accuracy. However, in this study, we focused on the role of astrocytes within a small network using a few-shot classification approach (Figs. 5, 6, and 7).

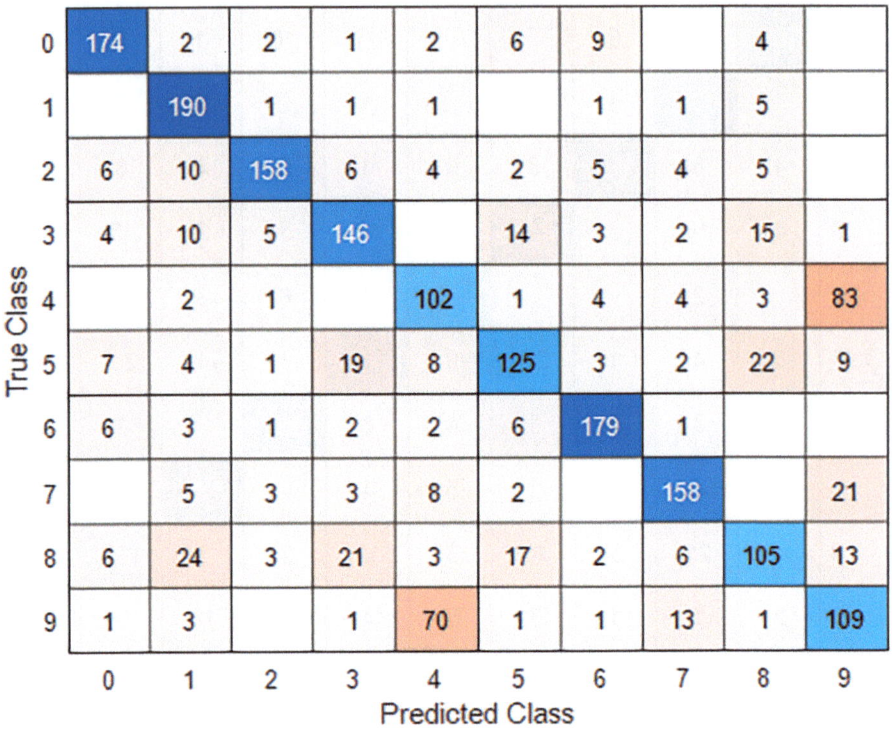

Fig. 6 Confusion matrix of the network with one astrocyte at each synapse

4 Conclusion

As mentioned earlier in this chapter, previous studies have highlighted the fundamental role of astrocytes [5–7], leading to the concept of the tripartite synapse, which consists of a presynaptic neuron, a postsynaptic neuron, and the surrounding astrocyte [38]. With the aim of demonstrating the role of astrocytes in neural networks, we employed spiking neural networks (SNNs), as they more accurately imitate the structure of the biological brain than traditional artificial neural networks. We extended our network by incorporating astrocytes, thereby creating a spiking neuron-astrocyte network (SNAN). To evaluate the importance of astrocytes in neural networks, we constructed such a network for learning and applied it to handwritten digit classification. Biological systems can form robust representations from just one or a few experiences; similarly, SNNs can effectively learn from limited data and generalize well, owing to their biologically inspired characteristics [14, 15]. Considering the points discussed above, few-shot learning was employed to evaluate the ability of the network. Two configurations of astrocyte placement and feedback mechanisms were examined. In the first configuration, each astrocyte was placed at the synapse between an excitatory and an inhibitory neuron, which increased accuracy because the astrocyte propagated synaptic events to other regions

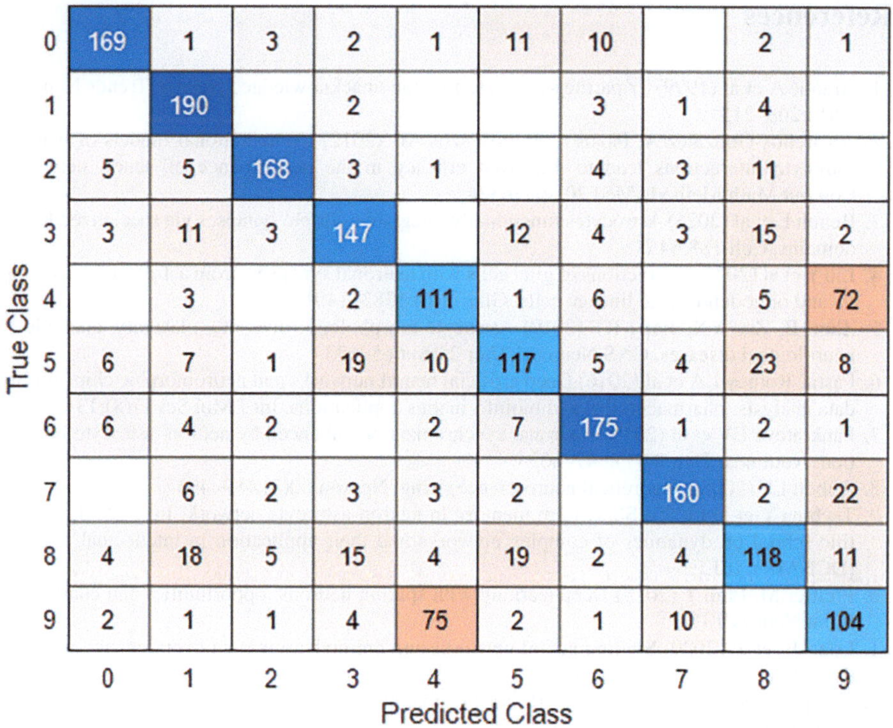

Fig. 7 Confusion matrix of the network with one astrocyte between two pairs of excitatory and inhibitory neurons

of the network. In other words, when the calcium concentration exceeds the threshold, the astrocyte sends feedback to neurons to regulate their synaptic strength. Some studies have shown that a single astrocyte can surround and influence numerous synapses from different pre- and postsynaptic neurons [5]. Inspired by this idea, we placed one astrocyte between two pairs of pre- and postsynaptic neurons in the second configuration. This design further improved performance, as astrocytic feedback and connectivity accelerated the convergence of the network to labels and enhanced its overall classification ability.

AI Use Disclosure Statement The authors used AI-based tools (ChatGPT, https://chat.openai.com, and Grammarly, https://www.grammarly.com) only to check grammar and do minor proofreading while writing this work.

These tools did not create the ideas, analysis, interpretation, or conclusions in the manuscript. The authors assert that all concepts, research, and scientific discourse were solely developed by them and accept full responsibility for the manuscript's content and integrity.

References

1. Araque A et al (1999) Tripartite synapses: glia, the unacknowledged partner. Trends Neurosci 22(5):208–215
2. Alvarellos-González A, Pazos A, Porto-Pazos AB (2012) Computational models of neuron-astrocyte interactions lead to improved efficacy in the performance of neural networks. Comput Math Methods Med 2012:476324
3. Benoit L et al (2025) Astrocytes functionally integrate multiple synapses via specialized leaflet domains. Cell 188:6453
4. Liu Y et al (2023) Interactions of glial cells with neuronal synapses, from astrocytes to microglia and oligodendrocyte lineage cells. Glia 71(6):1383–1401
5. Zhou B, Zuo YX, Jiang RT (2019) Astrocyte morphology: diversity, plasticity, and role in neurological diseases. CNS Neurosci Ther 25(6):665–673
6. Pastur-Romay LA et al (2016) Deep artificial neural networks and neuromorphic chips for big data analysis: pharmaceutical and bioinformatics applications. Int J Mol Sci 17(8):1313
7. Pankratova EV et al (2019) Neuronal synchronization enhanced by neuron–astrocyte interaction. Nonlinear Dyn 97(1):647–662
8. Abbott LF (2008) Theoretical neuroscience rising. Neuron 60(3):489–495
9. Tsybina Y et al (2020) Short-term memory in neuron-astrocyte network. In: 2020 4th scientific school on dynamics of complex networks and their application in intellectual robotics (DCNAIR). IEEE
10. Pfeiffer M, Pfeil T (2018) Deep learning with spiking neurons: opportunities and challenges. Front Neurosci 12
11. Lobo JL et al (2020) Spiking neural networks and online learning: an overview and perspectives. Neural Netw 121:88–100
12. Thorpe S, Delorme A, Van Rullen R (2001) Spike-based strategies for rapid processing. Neural Netw 14(6–7):715–725
13. Putra RVW, Shafique M (2024) NeuroNAS: enhancing efficiency of neuromorphic in-memory computing for intelligent mobile agents through hardware-aware spiking neural architecture search. arXiv preprint arXiv:2407.00641
14. Xu Q et al (2025) Self-cross feature based spiking neural networks for efficient few-shot learning. ArXiv:abs/2505.07921
15. Lake BM, Salakhutdinov R, Tenenbaum JB (2015) Human-level concept learning through probabilistic program induction. Science 350(6266):1332–1338
16. Orgel MG (1984) Epineurial versus perineurial repair of peripheral nerves. Clin Plast Surg 11(1):101–104
17. Chrysafides SM, Bordes SJ, Sharma S (2023) Physiology, resting potential. In: StatPearls [Internet]. StatPearls Publishing
18. Interspike interval (2010) Knowino, an encyclopedia. http://knowino.org/w/index.php?title=Interspike_interval&oldid=2936. Accessed 15 Feb 2015
19. Freeman MR, Rowitch DH (2013) Evolving concepts of gliogenesis: a look way back and ahead to the next 25 years. Neuron 80(3):613–623
20. Nikonenko I et al (2014) Glia and the synapse: plasticity and disease. Neural Plast 2014:246714
21. Nadkarni S, Jung P, Levine H (2008) Astrocytes optimize the synaptic transmission of information. PLoS Comput Biol 4(5):e1000088
22. Chakravarthy N et al (2009) Homeostasis of brain dynamics in epilepsy: a feedback control systems perspective of seizures. Ann Biomed Eng 37(3):565–585
23. Lee PR, Fields D (2009) Regulation of myelin genes implicated in psychiatric disorders by functional activity in axons. Front Neuroanat 3
24. Drachman DA (2005) Do we have brain to spare? Neurology 64(12):2004–2005
25. Niswender CM, Conn PJ (2010) Metabotropic glutamate receptors: physiology, pharmacology, and disease. Annu Rev Pharmacol Toxicol 50:295–322

26. Kandel ER, Schwartz JH, Jessell T (2000) Principles of neural science, 4th edn. McGraw-Hill Companies, Incorporated
27. Zonta M, Carmignoto G (2002) Calcium oscillations encoding neuron-to-astrocyte communication. J Physiol-Paris 96(3–4):193–198
28. Parpura V et al (1994) Glutamate-mediated astrocyte–neuron signalling. Nature 369(6483):744–747
29. Banerjee A et al (2016) Roles of presynaptic NMDA receptors in neurotransmission and plasticity. Trends Neurosci 39(1):26–39
30. Liu Q-S et al (2004) Astrocyte activation of presynaptic metabotropic glutamate receptors modulates hippocampal inhibitory synaptic transmission. Neuron Glia Biol 1(4):307–316
31. Sasaki T et al (2014) Astrocyte calcium signalling orchestrates neuronal synchronization in organotypic hippocampal slices. J Physiol 592(13):2771–2783
32. Diehl PU, Cook M (2015) Unsupervised learning of digit recognition using spike-timing-dependent plasticity. Front Comput Neurosci 9
33. Mazzoni A et al (2008) Encoding of naturalistic stimuli by local field potential spectra in networks of excitatory and inhibitory neurons. PLoS Comput Biol 4(12):e1000239
34. Stimberg B, Goodman D (2018). Brian 2 documentation, Release 2.2.1
35. Sjöström J, Gerstner W (2010) Spike-timing dependent plasticity. Scholarpedia 5(2):1362
36. Li Y, Rinzel J (1994) Equations for InsP3 receptor-mediated [Ca^{2+}]i oscillations derived from a detailed kinetic model: a Hodgkin-Huxley like formalism. J Theor Biol 166(4):461–473
37. Wade JJ et al (2011) Bidirectional coupling between astrocytes and neurons mediates learning and dynamic coordination in the brain: a multiple modeling approach. PLoS One 6(12):e29445
38. Perea G, Navarrete M, Araque A (2009) Tripartite synapses: astrocytes process and control synaptic information. Trends Neurosci 32(8):421–431

[illegible]

A Predictive Framework for Crack Width Assessment in GFRP-Reinforced Concrete Beams Using RSM

Alireza Arabshahi, Mohammad Meghdadian, Sima Rostami Aghoy, and Nima Gharaei-Moghaddam

Abstract GFRP bars have been considered a major alternative to traditional steel reinforcement in recent years owing to their favorable mechanical properties. FRP longitudinal reinforcements have been used more frequently in civil engineering projects during the past few decades, mainly because concrete structures often face harsh environmental conditions. Research performed on FRP-reinforced concrete, particularly beams, has found that two key serviceability aspects, i.e., deflection and cracking, can differ significantly from those of steel-reinforced beams. Accordingly, reliable relationships to make accurate predictions for these parameters are necessary. Crack width is an important measure to evaluate the post-cracking behavior of RC beams. The tensile stress in the longitudinal bars transfers to the surrounding concrete by an increase in the flexural moment, leading to crack propagation. Since various factors affect the crack width in GFRP-reinforced concrete beams, this study proposes an accurate relation for predicting crack width using the response surface methodology (RSM). A more precise predictive model compared to existing ones is proposed by applying polynomial functions. To assess the accuracy of the proposed model, a dataset of experimental test results from previously published studies is used. Moreover, a sensitivity analysis is performed using the RSM approach to evaluate the effect of different design parameters. The obtained results show that the proposed relation offers higher accuracy than the existing models and provides a more dependable estimation of crack width.

Keywords Crack width · Response surface method · Fiber-reinforced polymers Reinforced concrete

A. Arabshahi · M. Meghdadian · S. R. Aghoy · N. Gharaei-Moghaddam (✉)
Civil Engineering Department, Ferdowsi University of Mashhad, Mashhad, Iran
e-mail: alireza.arabshahi@mail.um.ac.ir; M.Meghdadian@mail.um.ac.ir; Sima.Rostami@mail.um.ac.ir; Nima.gharaeimoghaddam@mail.um.ac.ir

M. Mirrashid et al. (eds.), *Optimization and Adaptive Control Algorithms*, Emerging Trends in Mechatronics, https://doi.org/10.1007/978-981-95-8387-4_7

1 Introduction

The use of fiber-reinforced polymer (FRP) composites in civil engineering structures has grown considerably in the recent decade. This growth is both for new buildings as well as the strengthening and rehabilitation of existing structures. This increasing interest is mainly due to the positive mechanical and durability properties of FRPs, including but not limited to high tensile strength and elastic modulus, excellent corrosion resistance, lightweight, ease of installation and application, electromagnetic neutrality, and long-term performance and durability [1–5]. The mentioned benefits introduce FRP bars as a sustainable alternative to conventional steel reinforcement. These composite bars are especially favorable for construction and strengthening in corrosion-prone environments where chemical attacks are expected. As mentioned, in addition to their common applications in strengthening and retrofitting of structures (especially concrete structures), FRP composites have also been increasingly used as longitudinal reinforcement in newly constructed concrete structures [6–8]. Glass fiber-reinforced polymer (GFRP) has gained the most extensive adoption among the different FRP types, because of their advantages, such as cost-effectiveness, availability, and satisfactory mechanical performance. It should be mentioned that in spite of their benefits, FRP bars are inherently brittle and, compared to conventional steel reinforcement, exhibit weaker bond characteristics with concrete. These characteristics result in a lack of ductility and reduced energy absorption at ultimate conditions. However, recent studies proved that this shortcoming can be resolved partly by using hybrid steel-FRP reinforcements [9]. Moreover, the noncorrosive nature of FRP bars makes them a suitable option for use in structures subjected to aggressive marine, coastal, or de-icing environments, where steel reinforcement is prone to deterioration.

FRP bars demonstrate linear elastic behavior up to rupture. Accordingly, FRP-reinforced concrete (FRP-RC) members generally exhibit a brittle failure mode without the yielding phase typically observed in steel-reinforced members. As mentioned earlier, to mitigate this issue, hybrid reinforcement systems combining FRP and steel bars in the tensile zone have been suggested to improve ductility and post-cracking performance [8, 10]. These hybrid reinforcement strategies offer a desirable structural performance by simultaneously benefiting from the corrosion resistance of FRP and the ductile nature of steel. Because of the relatively weaker bond between FRP bars and concrete compared to that of steel bars, several studies have been conducted to experimentally, analytically, and/or numerically investigate and predict the behavior of GFRP-reinforced and hybrid RC beams [11–14]. These studies were mainly focused on serviceability aspects, including midspan deflection, cracking behavior, failure modes, crack width, and energy absorption capacity. Theriault and Benmokrane [15] performed one of the early research works in this field. They explored the effect of reinforcement ratio and concrete strength on the flexural behavior of CFRP-reinforced beams. In another study, Toutanji and Dong [16] proposed an empirical model based on experimental investigations to estimate crack width in GFRP-reinforced beams.

Further research has been performed to expand the understanding of the deflection and cracking performance of RC beams reinforced with FRP bars. In a major study, Kassem et al. [17] conducted full-scale experiments in order to evaluate deflection and flexural capacity. In other studies, Al-Sunna et al. and Mias et al. [18, 19] investigated both the short-term and long-term cracking of GFRP-reinforced beams subjected to cyclic loading. The findings of these investigators demonstrated inconsistencies between experimental results and predictions obtained from existing code-prescribed design relations. These unexpected findings highlighted the need for more accurate predictive serviceability models. In this regard, El-Nemr et al. [20] conducted experiments on normal- and high-strength concrete beams. They reported that bar diameter significantly affects crack width; however, it has a limited influence on deflection. In a similar study, Sturm et al. [21] evaluated CFRP-reinforced high-performance concrete beams. Again, they found noteworthy differences between experimental crack widths and those predicted by design provisions. Following these studies, El Refai et al. [22] proposed an equation with improved accuracy for predicting crack width in GFRP-reinforced concrete beams based on experimental findings.

In conjunction with the mentioned experimental studies, finite element modeling (FEM) has been used as a powerful tool to simulate the structural performance of FRP-reinforced beams under different loading conditions. For instance, Saleh et al. [23, 24] developed a three-dimensional numerical model to analyze GFRP-reinforced beams subjected to uniform loads. Using their developed model, they investigated the influence of key design parameters such as reinforcement ratio, beam geometry, and material properties on deflection, crack formation, and energy dissipation of FRP-reinforced beams. Once again, comparison of their numerical results against design code predictions revealed noticeable differences in flexural capacity and serviceability responses. Since GFRP materials possess a lower modulus of elasticity than steel, GFRP-reinforced beams typically experience larger deflections and wider cracks, particularly at service load levels [25, 26]. To address these discrepancies, several researchers have developed analytical and numerical models for predicting deflection and crack width in FRP-reinforced concrete beams. Kara and Ashour [27] introduced a numerical method to estimate flexural capacity and deflection, while Barris et al. [25] critically evaluated the accuracy of available predictive equations using results from multiple test programs. In spite of the reviewed advancements, the current design formulations often show considerable scatter when applied to beams with varying reinforcement ratios, concrete strengths, or bar surface distribution.

Since crack width is a key parameter governing the serviceability and durability of reinforced concrete beams, an accurate prediction model is essential. In this study, a new empirical relationship for predicting crack width in GFRP-reinforced concrete beams is developed using the response surface methodology (RSM). The model integrates experimental data from previous studies (as summarized in Sect. 1) to establish polynomial functions that capture the combined effects of influencing parameters. The predictive performance of the proposed model is validated and compared with existing formulations, and a sensitivity analysis is conducted to quantify the influence of key design variables. The ultimate goal is to provide a more accurate and reliable expression for estimating crack width in GFRP-reinforced beams, contributing to improved serviceability assessment and design practice.

2 Experimental Data on Crack Width

Many experimental studies have been performed during the past three decades in order to evaluate the performance of FRP-reinforced concrete (RC) beams. These studies have focused mainly on evaluating the flexural response, serviceability characteristics, and cracking performance of FRP-reinforced beams subjected to different loading and environmental conditions. The results of these experimental studies have been used for the development of predictive empirical equations for crack width in FRP-reinforced concrete beams. These experimental databases serve as the fundamental reference for validating analytical or numerical predictive models.

A review of the available literature provides a large number of experimental programs performed by different researchers. Each of these researchers investigated the influence of distinct design parameters such as beam geometry, reinforcement ratio, concrete compressive strength, and type or surface configuration of the GFRP bars. All in all, these experimental studies provide valuable insights into the influence of the mechanical and geometric parameters on the serviceability of GFRP-reinforced members. The main characteristics of the tested specimens in the most relevant investigations are summarized in Table 1.

The database presented in Table 1 was collected through careful screening and selection of experimental results. The adopted screening process ensures that only high-quality and comparable test results are included. For this purpose, a comprehensive literature review was first conducted to collect experimental studies that have been performed on simply supported beams reinforced with GFRP bars, which were tested under monotonic or static loading conditions. To maintain a homogeneous database, experimental results on beams reinforced with other FRP types or hybrid reinforcement systems and specimens subjected to cyclic or fatigue loading were excluded. Moreover, Table 1 only includes results from studies that reported complete details on geometric dimensions, reinforcement details, and mechanical properties of both concrete and FRP materials.

Several essential design parameters significantly affect the crack formation and propagation of the FRP-reinforced concrete beams. These effective design parameters include the height and width of the beam cross section (b and h), concrete cover, number and diameter of tensile rebars, compressive strength of concrete (f_c'), elastic modulus (E_f) and tensile strength (f_{fu}) of GFRP bars, and diameter of stirrups.

These geometrical properties of the section have a direct effect on the stiffness and moment of inertia of beams and consequently influence the distribution and spacing of cracks. Concrete cover is equal to the distance from the outer surface of

Table 1 Summary of previous experimental studies and parameters used in proposed crack width equations for GFRP-reinforced concrete beams

No.	Reference	b (mm)	h (mm)	Concrete cover (mm)	No. of tension rebar	Diameter of tension rebar (mm)	f_{co} (MPa)	E_f (MPa)	f_{fu} (MPa)	Diameter of stirrup (mm)
1	Barris et al. [25]	140	190	20	2	12	32.1	63,437	1321	8
2		140	190	20	2	12	32.1	63,437	1321	8
3		140	190	20	3	16	32.1	64,634	1015	8
4		160	190	40	2	12	32.1	63,437	1321	8
5		160	190	20	3	16	54.5	64,634	1015	8
6	Theriault and Benmokrane [15]	130	180	20	2	12.3	53.1	38,000	737	6
7		130	180	20	2	12.3	53.1	38,000	737	6
8		130	180	20	2	12.3	57.2	38,000	737	6
9		130	180	20	2	12.3	97.4	38,000	737	6
10		130	180	25	4	12.3	46.2	38,000	737	6
11		130	180	25	4	12.3	53.9	38,000	737	6
12		130	180	25	4	12.3	93.5	38,000	737	6

(continued)

Table 1 (continued)

No.	Reference	b (mm)	h (mm)	Concrete cover (mm)	No. of tension rebar	Diameter of tension rebar (mm)	f_{co} (MPa)	E_f (MPa)	f_{fu} (MPa)	Diameter of stirrup (mm)
13	El-Nemr et al. [20]	200	400	40	3	13	33.5	48,700	817	10
14		200	400	40	5	13	38.95	48,700	817	10
15		200	400	40	3	15	38.95	48,100	751	10
16		200	400	40	4	15	38.95	48,100	751	10
17		200	400	40	2	13	33.5	67,000	1639	10
18		200	400	40	2	15	29	69,300	1362	10
19		200	400	40	2	25	48.13	66,300	1132	10
20		200	400	40	3	20	48.13	52,500	1082	10
21		200	400	40	2	15	33.83	59,500	1245	10
22		200	400	40	2	25	33.8	60,300	906	10
23		200	400	40	2	20	38.95	47,600	728	10
24		200	400	40	3	20	42.1	47,600	728	10
25		200	400	40	2	22	38.95	46,400	693	10
26		200	400	40	2	25	48.13	53,200	666	10
27		200	400	25	6	15	33.5	48,100	751	10
28		200	400	25	5	15	29	69,300	1362	10
29		200	400	25	5	15	33.8	59,500	1245	10
30	Toutanji and Deng [16]	180	300	20	2	12.7	35	40,000	695	6
31		180	300	20	3	12.7	35	40,000	695	6
32		180	300	25	4	12.7	35	40,000	695	6
33	Goldston et al. [23]	100	150	15	2	9.53	40	55,600	1764	4
34		100	150	15	2	12.7	40	48,600	1605	4
35		100	150	15	2	9.53	80	55,600	1764	4
36		100	150	15	2	12.7	80	48,600	1605	4

No.	Reference	b (mm)	h (mm)	Concrete cover (mm)	No. of tension rebar	Diameter of tension rebar (mm)	f_{co} (MPa)	E_f (MPa)	f_{fu} (MPa)	Diameter of stirrup (mm)
37	Abdelkarim et al. [28]	200	300	25	2	12	35	65,000	1166	10
38		200	300	25	2	16	35	63,000	1122	10
39		200	300	25	2	20	35	69,000	1117	10
40		200	300	25	2	25	35	65,000	1340	10
41		200	300	25	2	16	65	63,000	1122	10
42		200	300	25	2	20	65	69,000	1117	10
43		200	300	25	2	25	65	65,000	1340	10
44		200	250	35	4	19	40.7	35,630	700	8
45	El Refai et al. [22]	230	300	40	3	12	40	50,000	1000	8
46		230	300	40	3	16	40	50,000	1000	8
47	Alsayed [29]	200	210	35	4	19	31	35,630	700	8
48		200	260	35	4	12.7	31	45,000	886	8
49		200	250	35	4	19	41	35,630	700	8
50	Al-Sunna et al. [18]	150	250	25	2	12.7	47.7	41,600	620	8
51		150	250	25	4	19.05	46.5	42,000	670	8
52		150	250	25	3	9.53	52.6	132,000	1320	8
53		150	250	25	2	9.53	47.7	42,800	665	8
54		150	250	25	3	6.35	55.4	133,000	1450	8
55		150	250	25	3	12.7	51.8	119,000	1475	8

(continued)

Table 1 (continued)

No.	Reference	b (mm)	h (mm)	Concrete cover (mm)	No. of tension rebar	Diameter of tension rebar (mm)	f_{co} (MPa)	E_f (MPa)	f_{fu} (MPa)	Diameter of stirrup (mm)
56	Saleh et al. [24]	100	150	25	2	6.35	55.4	37,500	732	4
57		100	150	25	2	9.53	55.4	55,600	1764	4
58		100	150	25	2	12.7	55.4	48,600	1605	4
59		100	150	25	2	6.35	70.8	37,500	732	4
60		100	150	25	2	9.53	70.8	55,600	1764	4
61		100	150	25	2	12.7	70.8	48,600	1605	4
62	Kassem et al. [17]	200	300	25	4	9.5	40.4	114,000	1506	10
63		200	300	25	4	9.5	40.4	114,000	1506	10
64		200	300	25	6	9.5	39.3	114,000	1506	10
65		200	300	25	8	9.5	39.3	114,000	1506	10
66		200	300	25	4	9	39.9	122,000	1988	10
67		200	300	25	4	9	39.9	122,000	1988	10
68		200	300	25	6	9	40.8	122,000	1988	10
69		200	300	25	8	9	40.8	122,000	1988	10
70		200	300	25	6	12.7	39.05	40,000	617	10
71		200	300	25	8	12.7	39.05	40,000	617	10
72		200	300	25	6	12	39.05	36,000	747	10
73		200	300	25	8	12	39.05	36,000	747	10
74		200	300	25	6	9.5	39.05	52,000	1800	10
75		200	300	25	8	9.5	39.05	52,000	1800	10

No.	Reference	b (mm)	h (mm)	Concrete cover (mm)	No. of tension rebar	Diameter of tension rebar (mm)	f_{co} (MPa)	E_f (MPa)	f_{fu} (MPa)	Diameter of stirrup (mm)
76	Kara and Ashour [27]	150	200	25	2	6	34	38,000	650	6
77		150	250	25	2	6	34	38,000	650	6
78		150	300	25	2	6	34	38,000	650	6
79		150	200	25	2	6	59	38,000	650	6
80		150	250	25	2	6	59	38,000	650	6
81		150	300	25	4	6	59	38,000	650	6
82	Erfan et al. [30]	150	300	25	2	8	50	21,000	482.24	8
83		150	300	25	2	10	50	38,000	636.37	8
84		150	300	25	2	8	60	21,000	482.24	8
85		150	300	25	2	12	50	37,500	745.64	8
86		150	300	25	2	12	60	37,500	745.64	8
87		150	300	25	2	10	60	38,000	636.37	8

the concrete to the center of the longitudinal GFRP bar. This distance controls the stress transfer mechanism and crack initiation at the tension face of the beams. The amount and configuration of the GFRP bars define the reinforcement ratio, which is one of the most important factors in determining both the crack width and spacing. It is widely known that a higher reinforcement ratio generally leads to narrower cracks but a larger number of crack lines. Compressive strength of concrete reflects the stiffness of the matrix. This parameter affects the concrete's ability to restrain crack widening. On the other hand, GFRP material properties control the stress-strain compatibility between the bar and the surrounding concrete. The lower elastic modulus of GFRP compared to steel typically leads to larger deflections and wider cracks at service loads. Finally, the diameter of stirrups affects the confinement of concrete and indirectly contributes to controlling crack width by restricting lateral expansion.

The inclusion of these parameters in the collected database confirms coverage of a wide and representative range of practical design conditions. It is noteworthy that the considered experiments were performed on simply supported beams with rectangular cross sections, which underwent four-point bending tests. This test setup allows for the formation of a constant-moment region between the loading points. In this constant-moment region, the crack initiation and propagation can be clearly observed and measured. Crack widths in the referenced studies were typically measured at midspan using either mechanical microscopes or digital displacement transducers, ensuring reliable and repeatable results.

The diversity of the collected dataset, which includes different ranges of the effective design parameters, provides a robust basis for the development of a generalized empirical model. Various researchers have proposed empirical equations to predict serviceability measures of FRP-reinforced beams under flexural loading by correlating these parameters with the observed crack widths. However, the existing scatter in the reported results necessitates a systematic statistical approach, such as the response surface methodology (RSM) used in this study, to improve the prediction accuracy.

It should be noted that, from the reported results in these studies, three different crack widths for different applied loads were recorded for each specimen, forming a database of 261 data points for the derivation of the proposed relation (80% of data points) and accuracy evaluation (20% of data points).

3 Available Equations for Crack Width in GFRP-Reinforced Concrete Beams

The process of cracking in reinforced concrete beams usually starts with microcrack formation within the tension zone. When the applied tensile strain reaches the ultimate elastic capacity of the concrete, the developed microcracks coalesce and form major visible cracks. These major cracks then propagate with increasing load. After cracking, the properties of the reinforcement and the bond interaction between the GFRP bars and the concrete matrix govern the flexural behavior of the beam. In this range of behavior, three major serviceability parameters, namely the number of cracks, their average width, and the spacing between adjacent cracks, describe the behavior of the beam and significantly affect its long-term durability.

During the past few years, various researchers and professional design bodies have attempted to better understand the crack development mechanisms in GFRP-reinforced concrete beams. Their target has been to quantify the cracking behavior and provide reliable and practical predictive tools. For this purpose, different analytical, semiempirical, and numerical models have been suggested. Most of the previous predictive models were developed and calibrated based on detailed and comprehensive experimental datasets. Some of the most widely used formulations are presented in Table 2. Each of these models was developed considering specific assumptions regarding the bond between the reinforcement and concrete, the stiffness of the bars, and the mechanism of stress distribution after cracking.

Most of these relationships define the average crack width (w) as a function of the tensile stress in the reinforcement (f_f), elastic modulus of the GFRP bar (E_f), and geometric parameters that describe the beam's configuration and reinforcement layout. An important parameter in these relationships is the bond factor (k_b). This parameter characterizes the bond efficiency between the FRP bar and concrete and is highly dependent on the surface texture, type of resin, and manufacturing process of the GFRP bars. Because the precise value of the bond factor cannot be directly determined, most design codes recommend using a range of values, as summarized in Table 2. In these equations, h and d stands for the total depth and the effective depth of the beam, while ϕ and s represent the diameter and the distance between FRP bars. The other parameters used in these relations are computed using the subsequent equations:

Table 2 Existing equations for crack width prediction in GFRP-reinforced concrete beams

No.	Model	Equations
1	ACI 440.1R-06 [31]; CAN/CSA S6-06 [32]	$w = 2\frac{f_f}{E_f}\beta k_b\sqrt{d_c^2 + \left(\frac{s}{2}\right)^2}$ $k_b = 0.84{\sim}1.1$
2	ISIS [33]	$w = 2.2\frac{f_f}{E_f}\beta k_b\sqrt[3]{d_c A}$ $k_b = 0.68{\sim}0.89$
3	JSCE [34]	$w = k_b\left(4c + 0.5(s - \phi)\right)\frac{f_f}{E_f}$ $k_b = 0.82{\sim}1.07$
4	Toutanji and Saafi [35]	$w = 0.002\frac{f_f}{E_f\sqrt{\rho_f}}\beta\sqrt[3]{d_c A}$

$$m = \frac{E_f}{E_c} \tag{1}$$

$$\rho_f = \frac{A_f}{bd} \tag{2}$$

$$k = \sqrt{(\rho_f m)^2 + 2\rho_f m} - \rho_f m \tag{3}$$

$$\beta = \frac{h - kd}{d(1-k)} \tag{4}$$

$$d_c = h - d \tag{5}$$

$$A = \frac{2(h-d)b}{n} \tag{6}$$

$$f_f = \min\left(\frac{M_{applied}}{A_f d\left(1 - \frac{k}{3}\right)}, 0.3 f_{fu} \right) \tag{7}$$

where n and f_{fu} are the number and the ultimate tensile strength of FRP bars, and A_f stands for the total area of FRP bars.

4 Proposed Equation for Crack Width in GFRP-Reinforced Concrete Beams

In this section, a new, more accurate predictive relation for estimating crack width in GFRP-reinforced concrete beams will be formulated using the response surface methodology (RSM). RSM is a powerful and robust framework for mathematical and statistical analyses of systems that are controlled by the simultaneous influence of different effective parameters. The primary objective of RSM is to identify the most influential parameters and their interactive effects and then develop a functional relationship between the effective inputs and the required output, which is the crack width in this study. RSM utilizes polynomial representations to model complex nonlinear engineering phenomena that cannot be easily captured using simple regression methods.

In this study, the considered input parameters include the mechanical properties, such as the level of tensile stress in GFRP bars and their modulus of elasticity, as well as geometric parameters that characterize the cross-sectional properties of the beam. By using a systematic fitting process on the collected experimental data from previous literature, different polynomial representations (linear, quadratic, cubic, quartic, and fifth order) will be examined for predicting crack width as a function of input parameters. The most suitable representation will be selected based on statistical measures.

The calculated measures and indicators for different polynomial forms are presented in Table 3. These measures, which collectively describe the goodness of fit for the model, include the sum of squares of errors, adjusted and predicted R^2, and F-statistics. It is evident in the table that the cubic form provides the best balanced statistical performance, which is a combination of as high a coefficient of determination (adjusted and predicted R^2) as possible and an acceptable prediction accuracy (lowest error measures). In the case of R^2 values, however, the quartic and fifth-order models demonstrate slightly higher adjusted values, but their predicted values drop sharply. This is an indicator that these higher-order models tend to overfit the input data and consequently will surely lack the generalization capability.

This finding is consistent with the common practice in modeling using response surfaces, where cubic models are usually preferred due to their ability to acceptably capture nonlinear variations without producing complex and spurious correlations. The results of the P-values also support the selection of the cubic model. Based on the attained results ($P = 0.0037$), the inclusion of the higher-order term in the model significantly improves the performance of the model in comparison to the lower-order polynomial models.

Based on the performed RSM analysis, the following relationship was selected as the most accurate and suitable predictive model for crack width (w) of GFRP-reinforced concrete beams:

$$W = \left(-0.0092 + 0.003688 f_{fu} + 0.000013A - 1.84\times10^{-6} f_f^2 - 3.13\times10^{-16} E_f^3\right)^2 \quad (8)$$

where f_{fu} represents the available tensile stress of the GFRP bars and A stands for a parameter that represents the geometric characteristics of the section, as defined by Eqs. (7) and (6), respectively. Moreover, E_f is the elastic modulus of the reinforcement. It should be noted that the inclusion of both linear and nonlinear terms enables the model to capture the gradual transition from microcracking to stabilized crack formation. This is the observed performance in the experiments. It is worth

Table 3 Comparison of polynomial models for fitting experimental data in crack width prediction

Source	Sequential P-value	Sum of square	Adjusted R^2	Predicted R^2	Mean square	P-value	F-value
Linear	<0.0001	11.04	0.371	0.358	0.103	<0.0001	3.07
2FI	0.304	10.79	0.373	0.352	0.103	<0.0001	3.09
Quadratic	0.3379	10.56	0.374	0.344	0.104	<0.0001	3.11
Cubic	0.0037	8.83	0.423	0.374	0.097	<0.0001	2.89
Quartic	0.0453	7.25	0.455	−0.185	0.095	<0.0001	2.84
Fifth	0.0008	4.6	0.535	−20.91	0.083	<0.0001	2.49

mentioning that the squared form of the final expression was selected to guarantee positive-definite predictions. This feature will enhance the model's stability when applied to a broad range of input variables. This makes the proposed relationship particularly useful for practical purposes, because this generalized nature allows the model to provide accurate predictions within and slightly beyond the range of experimental data used for its calibration.

To further validate the selected model, an analysis of variance (ANOVA) has been performed, and its results are summarized in Table 4. The ANOVA is usually used to quantify the contribution of each individual term to the final prediction and characterize the ability of each term in covering the total variation observed in the dataset. The results of ANOVA are presented in Table 4.

As can be seen, an overall F-value of 37.76 and a corresponding P-value of less than 0.0001 prove that the model itself is highly significant. This confirms that the selected parameters effectively simulate and capture the observed variation in crack width.

Among the considered effective variables in the model, the level of tensile stress in the GFRP reinforcement (f_f) demonstrates the strongest statistical significance ($F = 96.55$). The geometric representative parameter (A) and the cubic term of the elastic modulus (E_f^3) are in the next spots. This observation is in complete agreement with the physical nature of GFRP-reinforced concrete cracking. It is known that the tensile capacity of the reinforcement governs the strain distribution and crack formation along the beam length. Moreover, the higher-order terms in the proposed formulation confirm the nonlinear nature of the problem.

The residual and lack-of-fit components, reported in Table 4, are further indications of the model's accuracy. The attained significant lack of fit ($P < 0.0001$) is acceptable in the context of RSM applications. This reflects the inherent natural scatter in the multisource experimental datasets. Moreover, the relatively low value of pure error (0.0336) shows that the measurement errors within the source data were minimal. This finding somehow confirms the overall confidence of the developed regression model.

In the following section, the capability and accuracy of the proposed model in predicting the crack width of GFRP-reinforced concrete beams are evaluated and compared with other existing models, using the equations given in Table 2.

Table 4 Analysis of variance (ANOVA) for the proposed crack width model

Source model	Sum of square	Mean square	F-value	P-value	
Model	9.82	2.45	37.76	<0.0001	Significant
f_f	6.27	6.27	96.55	<0.0001	
A	1.69	1.69	25.97	<0.0001	
A^2	0.319	0.319	4.91	0.0277	
E_f^3	2.7	2.7	41.49	<0.0001	
Residual	13.52	0.065			
Lack of fit	10.09	0.0952	2.83	<0.0001	Significant
Pure error	3.43	0.0336			
Cor total	23.34				

5 Accuracy Assessment of the Proposed Equation

This section evaluates and compares the proposed predictive model's accuracy in predicting crack width to assess its robustness and show its superiority over the other existing models. For this purpose, 20% of the collected experimental data, summarized in Table 1, which was not used in the development of the RSM relation, is used.

For this purpose, the theoretical predictions of the predictive models are compared with the corresponding experimental outcomes. Four different statistical error indices are used to evaluate the accuracy of the predictions. These indices, which are commonly used in accuracy estimation, are defined as follows:

- Root mean square error:

$$\mathrm{RMSE} = \sqrt{\frac{\sum_{i=1}^{m}\left(\frac{X_{the} - X_{\exp}}{X_{\exp}}\right)^2}{n}} \tag{9}$$

- This index determines the average prediction error while giving more weight to large deviations.
- Mean absolute error:

$$\mathrm{MAE} = \frac{\sum_{i=1}^{m}\left|\frac{X_{the} - X_{\exp}}{X_{\exp}}\right|}{n} \tag{10}$$

- This index gives an intuitive measure of the average absolute error between predicted and experimental values.
- Integral absolute error:

$$\mathrm{IAE} = \frac{\left(\sum_{i=1}^{m}\sqrt{\left|\left(X_{the} - X_{\exp}\right)^2\right|}\right)}{\sum_{i=1}^{n}X_{\exp}} \tag{11}$$

- IAE criterion determines the cumulative normalized error with respect to the total experimental outcomes. This index offers a perspective of the total accuracy over all data points.
- Coefficient of determination:

$$R^2 = \left[\frac{\left(\sum_{i=1}^{m}\left(X_{the} - \overline{X}_{the}\right)\left(X_{\exp} - \overline{X}_{\exp}\right)\right)}{\left(\sqrt{\sum_{i=1}^{m}\left(X_{the} - \overline{X}_{the}\right)^2\sum_{i=1}^{n}\left(X_{\exp} - \overline{X}_{\exp}\right)^2}\right)}\right]^2 \tag{12}$$

This index specifies the degree of linear correlation between theoretical predictions and experimental results. A value close to 1.0 signifies a strong positive linear correlation, which is an indication of model accuracy. In these relationships, X represents a general response quantity; X_{the} and X_{exp}, respectively, denote the predicted theoretical value and the experimental results; and $\bar{X}_{the}$ and $\bar{X}_{\exp}$ are the corresponding mean values, respectively. m also stands for the number of data points.

The first three indices (RMSE, MAE, and IAE) are error measures, and their smaller values signify the model's higher accuracy. On the other hand, the coefficient of determination (R^2) serves as a model fidelity measure.

The computed indices for different relations, including the proposed model as well as equations derived from ACI, CSA, and JSCE codes, are presented in Fig. 1. It is evident that the suggested RMS-based model has lower RMSE and MAE error indices. This means that the proposed model provides better accuracy and superior generalization.

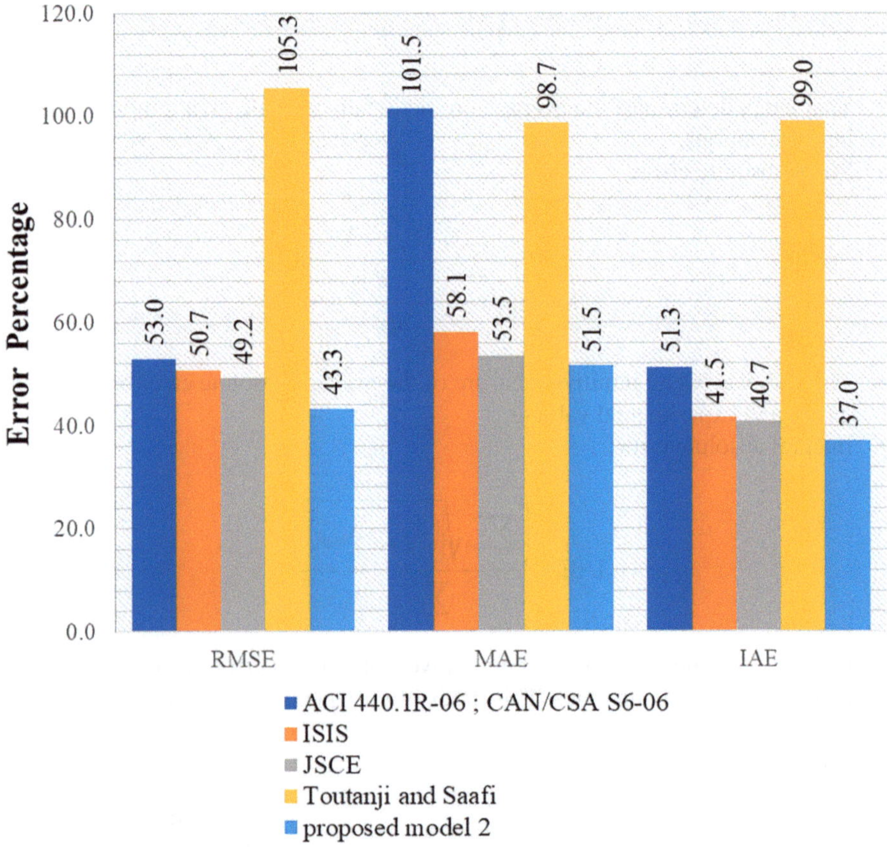

Fig. 1 Comparison of prediction errors for the proposed equation with experimental data and existing crack width models

The scatter and correlation between the predicted crack width values and the experimental results are demonstrated in Fig. 2. It can be seen that the data points corresponding to the suggested formulation are better aligned with the line of perfect correlation (R^2 = 1). This shows the acceptable agreement between the predicted values by the suggested model and the experimental results. It is also evident from the scatter diagrams that the models proposed by different codes such as ISIS, ACI, and JSCE somehow overestimate the crack width of GFRP-reinforced concrete beams. They also show greater scatter compared to the proposed model.

The performed analysis in this section overall confirms that the suggested formulation is more accurate and reliable compared to the other existing predictive

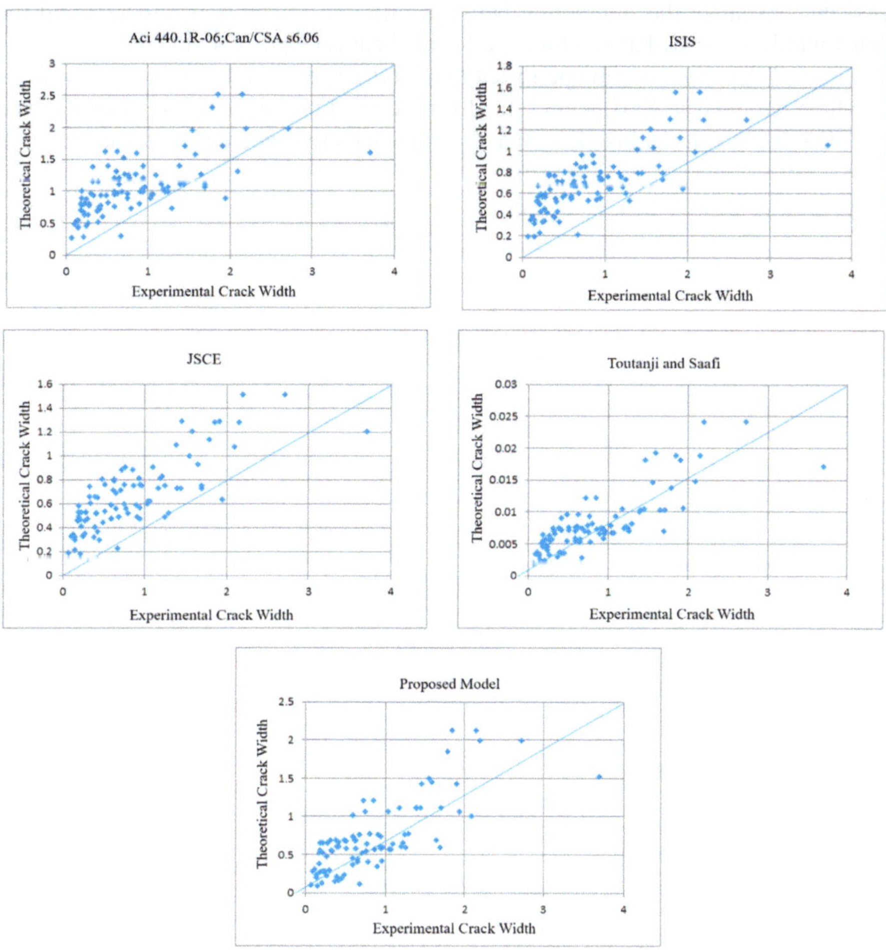

Fig. 2 Correlation plots between experimental crack widths and predicted values from the proposed model and existing equations

relationships for predicting the crack width of GFRP-reinforced concrete beams. This higher accuracy results from using statistically optimized coefficients derived from calibration by RSM methodology.

6 Parametric Study on Crack Width in GFRP-Reinforced Concrete Beams

In this section, a comprehensive parametric study is conducted to gain a deeper understanding of how various influential parameters affect the crack width of GFRP-reinforced concrete beams. For this purpose, a series of RSM plots are generated that visualize the interaction between different effective parameters and the crack width of the GFRP-reinforced concrete beams (Figs. 3, 4, and 5).

Figure 3 demonstrates the perturbation plot derived from the suggested formula, in which parameters A, B, and C stand for the level of tensile stress in GFRP bars (f_f), their elastic modulus (E_f), and the representative geometric parameter (A),

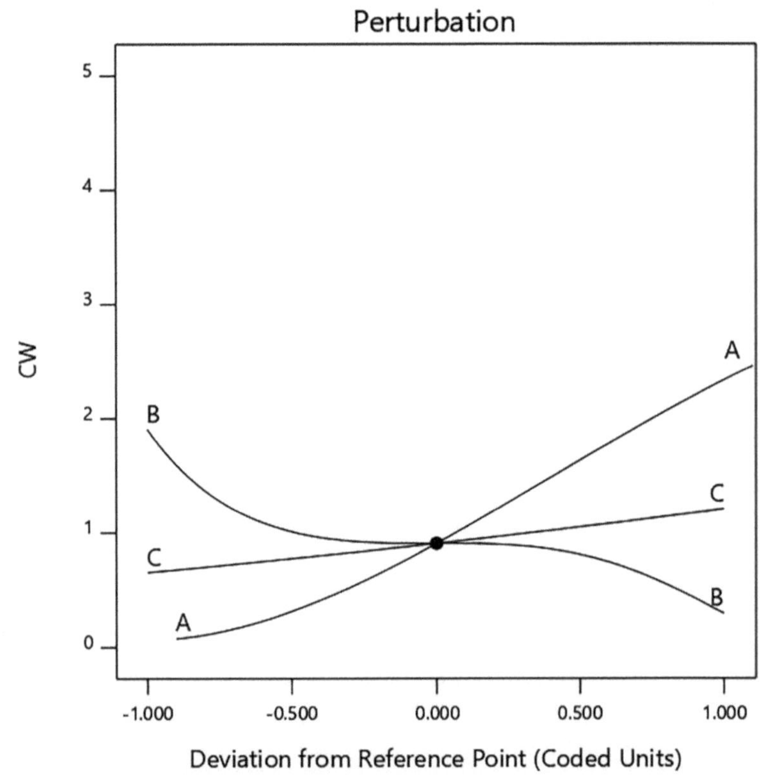

Fig. 3 Perturbation plot of the proposed relation showing the effect on the crack width response

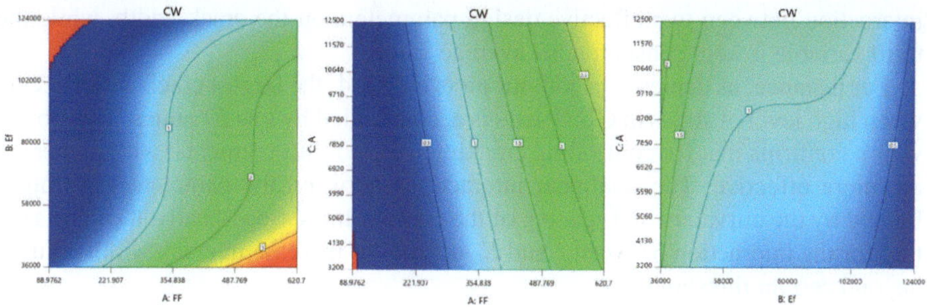

Fig. 4 Two-dimensional contour plots illustrating the combined influence of two parameters on crack width while holding the third parameter at its mean value

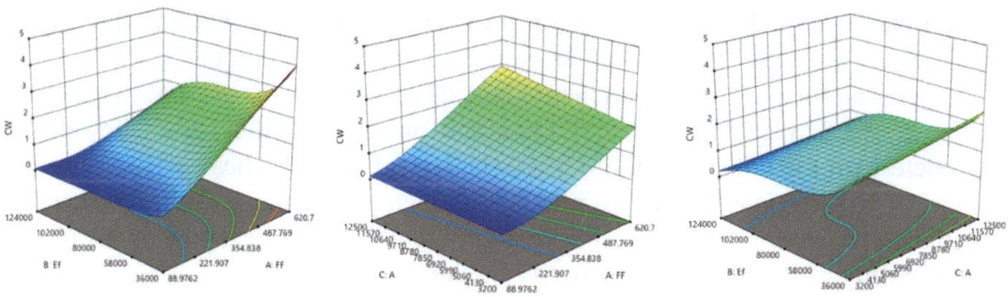

Fig. 5 Three-dimensional response surface plots depicting the joint effect of two parameters on crack width prediction

respectively. In this figure, each curve shows the variation of crack width with respect to one parameter, while the other two are kept constant at their mean values. This is a measure to evaluate the sensitivity of crack width to different design parameters.

As can be seen in the figure, there is an increase in f_f while E_f and A are kept constant, leading to a noticeable increase in crack width. This shows a positive correlation between the level of tensile stress in GFRP bars and the crack width. This trend can be justified by the fact that higher f_f values mean higher tensile stress in the surrounding concrete, which increases crack width. In another case, when f_f and A are kept constant, there is a negative relationship between modulus of elasticity (E_f) and crack width. This observation can be interpreted in the way that a higher modulus of elasticity means a higher stiffness of the reinforcement. Therefore, the strain compatibility effects are reduced, and consequently, the extent of crack formation is limited. Finally, the perturbation of the geometric parameter (A) shows a positive correlation. However, as it is evident, crack width is less sensitive to this parameter.

In order to better visualize the combined effect of the effective design parameters on the crack width of GFRP-reinforced concrete beams, two-dimensional contour plots are presented in Fig. 4. Moreover, for better clarity, three-dimensional response surface plots are also depicted in Fig. 5. These diagrams clearly illustrate the interaction between two effective design parameters on the crack width, while the third variable is kept constant at its average value.

For instance, as depicted in contour plots in Fig. 4, the largest crack width occurs when high level of stress in GFRP bars is accompanied by low E_f values. This observation confirms that for controlling cracking, stiffer reinforcements (with higher E_f) are more effective. The nonlinear interactions between the parameters are highlighted by the curvature and spacing of the contour lines. Specifically, it can be seen that for the interaction between f_f and E_f, the rate of increase in crack width reduces beyond certain thresholds.

The 3D surface plots depicted in Fig. 5 provide a better image of the interaction between effective design parameters. In these plots, instead of linear planes, smooth curved surfaces better demonstrate the variation trends. This figure clearly illustrates the evolution of crack width with the simultaneous variation of two effective design parameters. This demonstration provides valuable insights into the relative significance of each factor, as well as their combined impact on crack width.

In general, the performed parametric analysis ascertains the accuracy of the proposed model in capturing the individual and interactive influence of key design parameters on the crack width of GFRP-reinforced concrete beams. The observed trends in the response surfaces are in complete agreement with the expected mechanical behavior of GFRP-reinforced concrete beams. This agreement supports the reliability of the proposed predictive model.

7 Conclusion

In recent years, the application of FRP composites for strengthening and rehabilitation of structures, as well as building new structures, has grown significantly. This rapid growth in their application is owed to their positive properties, including high corrosion resistance, electromagnetic neutrality, high strength and stiffness, and long-term durability. These advantages make FRP bars a fascinating alternative to conventional steel bars, especially in harsh and aggressive environmental conditions. Among different types of FRP bars, glass fiber-reinforced polymer (GFRP) bars provide a cost-effective option for practical applications that gives a balance between performance and affordability.

In this study, a more accurate and reliable predictive model for crack width in GFRP-reinforced concrete beams was proposed using the response surface methodology (RSM). The suggested model considers the most influential design parameters, including the tensile stress in GFRP bars and their elastic modulus, as well as geometric and material properties of the concrete section. The adopted methodology allowed a systematic sensitivity evaluation of the effective parameters. This approach provides a clear understanding of the different parameter contributions to the cracking of beams. The proposed model was validated using a comprehensive experimental dataset. The attained results proved that the proposed RSM-based formulation provides superior accuracy compared to existing code-based predictive models, such as those suggested by ACI, CSA, and JSCE. The suggested formulation shows significantly lower errors and more consistent estimations of crack width.

The overall findings of this study confirm that the proposed RSM-based model provides a valuable and robust predictive tool for the design and analysis of GFRP-reinforced concrete beams. The application of the proposed model not only improves prediction accuracy but also provides useful insights into the effects of different design parameters, ensuring safety and serviceability in next-generation durable structures.

AI Use Disclosure Statement The authors declare that they have not used any generative AI tools in the preparation of the scientific content and analysis of this manuscript. Only for the purpose of final language proofreading and minor grammatical modifications were Grammarly and its integrated AI assistant employed.

References

1. Bouziadi F et al (2020) Finite element modeling of creep behavior of FRP-externally strengthened reinforced concrete beams. Eng Struct 204:109908
2. Arabshahi A, Gharaei-Moghaddam N, Tavakkolizadeh M (2019) Proposition of new applicable strength models for concrete columns confined with fiber reinforced polymers. SN Appl Sci 1(12):1677
3. Arabshahi A, Gharaei-Moghaddam N, Tavakkolizadeh M (2020) Development of applicable design models for concrete columns confined with aramid fiber reinforced polymer using Multi-Expression Programming. Structure 23:225–244
4. Gharaei-Moghaddam N, Arabshahi A, Tavakkolizadeh M (2023) Predictive models for the peak stress and ultimate strain of FRP confined concrete cylinders with inclined fiber orientations. Results Eng 18:101044
5. Jafar AA et al (2025) The influence of small fiber deviation on the behavior of CFRP confined concrete cylinders: an experimental study. Multiscale Multidiscip Model Exp Des 8(9):412
6. Rostami Aghouy S, Sabzi J, Esfahani MR (2025) The effect of long-term and short-term loading on the flexural behavior of RC beams strengthened with the NSM method. Case Stud Constr Mater 22:e04191
7. Sabzi J et al (2020) Effect of concrete strength and longitudinal reinforcement arrangement on the performance of reinforced concrete beams strengthened using EBR and EBROG methods. Eng Struct 205:110072
8. Nanni A (2003) North American design guidelines for concrete reinforcement and strengthening using FRP: principles, applications and unresolved issues. Constr Build Mater 17(6):439–446
9. Tavakol M, Haji Kazemi H (2025) Comparative assessment of concrete columns reinforced with hybrid steel-GFRP, GFRP, and steel bars under cyclic lateral loading. Structure 71:108034
10. Acciai A et al (2016) Experimental response of FRP reinforced members without transverse reinforcement: failure modes and design issues. Compos Part B 89:397–407
11. Alam MS, Hussein A (2017) Relationship between the shear capacity and the flexural cracking load of FRP reinforced concrete beams. Constr Build Mater 154:819–828
12. Cai J, Pan J, Zhou X (2017) Flexural behavior of basalt FRP reinforced ECC and concrete beams. Constr Build Mater 142:423–430
13. Dhahir MK, Nadir W (2018) A compression field based model to assess the shear strength of concrete beams reinforced with longitudinal FRP bars. Constr Build Mater 191:736–751

14. Dong Z et al (2019) Behaviors of hybrid beams composed of seawater sea-sand concrete (SWSSC) and a prefabricated UHPC shell reinforced with FRP bars. Constr Build Mater 213:32–42
15. Thériault M, Benmokrane B (1998) Effects of FRP reinforcement ratio and concrete strength on flexural behavior of concrete beams. J Compos Constr 2(1):7–16
16. Toutanji H, Deng Y (2003) Deflection and crack-width prediction of concrete beams reinforced with glass FRP rods. Constr Build Mater 17(1):69–74
17. Kassem C, Farghaly Ahmed S, Benmokrane B (2011) Evaluation of flexural behavior and serviceability performance of concrete beams reinforced with FRP bars. J Compos Constr 15(5):682–695
18. Al-Sunna R et al (2012) Deflection behaviour of FRP reinforced concrete beams and slabs: an experimental investigation. Compos Part B 43(5):2125–2134
19. Miàs C et al (2015) Short and long-term cracking behaviour of GFRP reinforced concrete beams. Compos Part B 77:223–231
20. El-Nemr A, Ahmed EA, Benmokrane B (2013) Flexural behavior and serviceability of normal- and high-strength concrete beams reinforced with glass fiber-reinforced polymer bars. ACI Struct J 110(6)
21. Sturm AB et al (2020) Flexural performance of pretensioned ultra-high performance fibre reinforced concrete beams with CFRP tendons. Compos Struct 243:112223
22. El Refai A, Abed F, Al-Rahmani A (2015) Structural performance and serviceability of concrete beams reinforced with hybrid (GFRP and steel) bars. Constr Build Mater 96:518–529
23. Saleh Z et al (2019) Flexural design of GFRP bar reinforced concrete beams: an appraisal of code recommendations. J Build Eng 25:100794
24. Saleh Z et al (2019) Numerical investigations on the flexural behavior of GFRP-RC beams under monotonic loads. Structure 20:255–267
25. Barris C et al (2013) Cracking and deflections in GFRP RC beams: an experimental study. Compos Part B 55:580–590
26. D'Antino T, Pisani MA (2018) Influence of sustained stress on the durability of glass FRP reinforcing bars. Constr Build Mater 187:474–486
27. Kara IF, Ashour AF (2012) Flexural performance of FRP reinforced concrete beams. Compos Struct 94(5):1616–1625
28. Abdelkarim OI et al (2019) Flexural strength and serviceability evaluation of concrete beams reinforced with deformed GFRP bars. Eng Struct 186:282–296
29. Alsayed SH, Al-Salloum YA, Almusallam TH (2000) Performance of glass fiber reinforced plastic bars as a reinforcing material for concrete structures. Compos Part B 31(6):555–567
30. Erfan AM et al (2020) The flexural behavior of nano concrete and high strength concrete using GFRP. Constr Build Mater 247:118664
31. 440, A.C.I.C (2006) Guide for the design and construction of structural concrete reinforced with FRP bars: ACI 440.1 R-06. American Concrete Institute
32. Code CJCSA (2006) Canada, CAN/CSA S6-06. 734
33. ISIS (2007) 3: Reinforcing concrete structures with fibre reinforced polymers. ISIS, pp 449–458
34. Teng J (2001) JSCE recommendations for upgrading of concrete structures with use of continuous fiber sheets. In: FRP composites in civil engineering: proceedings of the international conference on FRP composites in civil engineering, 12–15 December 2001, Hong Kong, China. Taylor & Francis
35. Toutanji HA, Saafi M (2000) Flexural behavior of concrete beams reinforced with glass fiber-reinforced polymer (GFRP) bars. ACI Struct J 97(5):712–719

Sliding Mode Control Optimized with PSO for Precise Trajectory Tracking of a Two-Axis Gimbal

Hamed Khodadadi

Abstract A two-axis gimbal system is designed utilizing two perpendicular single-axis gimbal systems. The optimization objectives for controlling this system are to obtain the optimal rotation paths around the *y*- and *z*-axes, with the requirement of covering all lines of sight (LOS). To achieve these objectives, the two control inputs, which are DC motor voltages, are used to adjust the angles of the two axes, which serve as the system's outputs, efficiently tracking the reference points. In this chapter, a nonlinear, robust approach, a sliding mode controller (SMC), is proposed for precise path tracking for two-axis gimbal movement. The objectives of this system are to minimize the time required to achieve the desired rotation path and to reduce the system's steady-state error. To optimize performance, the concept of SMC is combined with the particle swarm optimization (PSO) algorithm. In other words, by integrating SMC into the PSO algorithm, the hybrid approach benefits from SMC's ability to drive system dynamics robustly and rapidly toward desired states while retaining the population-based global search capability of PSO. This approach leads to a more efficient, stable, and disturbance-resistant system in complex problem spaces. Simulation results show that the proposed controller outperforms both the PID and conventional sliding mode controllers when tracking various command signals. Additionally, in the presence of system uncertainty, the optimal sliding mode controller is the most effective for achieving the best performance. This confirms the robustness of this controller in dynamic operating conditions.

Keywords Two-axis gimbal · Sliding mode controller · Particle swarm optimization · Optimal sliding mode controller

---

H. Khodadadi (✉)
Department of Electrical Engineering, Kho.C., Islamic Azad University, Khomeinishahr, Iran
e-mail: Hamed.khodadadi@iau.ac.ir

M. Mirrashid et al. (eds.), *Optimization and Adaptive Control Algorithms*, Emerging Trends in Mechatronics, https://doi.org/10.1007/978-981-95-8387-4_8

1 Introduction

For the improvement of the position of the camera on a drone to precise control of the robotic arm, dual-axis gimbals are used in various fields such as aerospace, robotics, and imaging. Their ability to maintain precise rotation and orientation is very important for the stability and control of such systems. Controlling a satellite's attitude is accomplished with the help of various actuators. These actuators consist of reaction wheels, control moment gyros, reaction control thrusters, and magnetic torque rods. Reaction control thrusters are used extensively due to their ability to generate large torques required for satellite attitude control in the presence of significant orbital changes [1]. On the other hand, despite the simplicity of the mechanical configuration, presence of nonlinear dynamics, and disturbances and uncertainty are challenges for these systems.

The level of torque a thruster can apply around a satellite axis depends on the thrust level and the length of the moment arm around that axis. This means that the effective use of thrusters is fundamentally dependent on their location on the satellite and their mounting angle relative to the satellite's body axes. In some cases, different torques may be required around the three main body axes of the satellite, which means the position and orientation of the thrusters must be carefully considered before the final physical setup of the system [2]. Thruster arrangements are influenced by factors such as the position of optical sensors, solar panels, antennas, etc.

Controlling motion in two dimensions and around two perpendicular axes is necessary in some applications. For this purpose, a two-axis gimbal system has been developed, which is formed by combining two single-axis gimbal systems placed perpendicularly to each other. These gimbals are designed to allow for easy rotation around each axis. The features of a two-axis gimbal system include 360° and 180° for the horizontal and vertical rotations, respectively, which cover the entire spherical space around it [3–8].

In real-world applications such as stabilizing a drone in high winds or controlling a robotic arm in irregular movements, PID controllers have been proposed as one of the conventional controllers to maintain control performance. However, the limitations of these controllers have led to the study of more advanced control approaches that can operate in dynamic and uncertain environments. On a satellite equipped with a two-gimbal mechanism, each of its three degrees of freedom rotation requires independent motor control. The system must produce precise motor signals to maintain stable trajectory tracking [1, 2]. The control input required for accurate satellite position stabilization is very affected by two important factors: the mass of the entire spacecraft and how its parts are arranged. The system parameters can change significantly during operations due to factors such as fuel consumption, payload deployment, or structural flexing. In these circumstances, the robust control approaches are suggested to maintain precise attitude control despite these changes [2–4]. Therefore, researchers have investigated these fields to improve spacecraft trajectory tracking [1]. A two-axis gimbal is characterized by a simple mechanical structure. It consists of a series of interconnected rings that create axes with a right angle around the connected piece, maintaining stability for that piece. This principle allows for stabilization and directional control of payloads such as cameras and sensors on satellite platforms [1–7]. The satellite can rotate in three different directions, with one engine required for each degree of freedom to rotate. The control structure applies a signal to the engine segment connected to the satellite and tracks the path of movement in a stable manner [1, 2]. The amount of control input in order to rotate in different directions depends on the weight and length of the arm attached to the satellite. Thus, the satellite rotation system can be assumed to be an indeterminate system that can be solved using resistant structures for this challenge [2–4]. Various researchers have evaluated and examined the effects of spacecraft stimuli and operators to track a specific path. In this study, the principles of optimizing movement in the desired direction were considered [1]. In [8], PSO has been suggested to improve the speed and stability of dual-gimbal control systems. Also, an SMC has been applied for the robustness of the system against uncertainties [1, 9–11].

The study [1] introduces a nonlinear fast terminal SMC (FTSMC) combined with a state-dependent Riccati equation (SDRE) observer for controlling a two-axis gimbal system under external disturbances. This study achieved finite-time stability and superior performance compared to existing methods. The demonstrated robustness against external disturbances provides applicability to various challenging operational environments, including aerospace systems and precision manufacturing equipment.

Furthermore, researchers in [12–14] have investigated H∞ control for greater disturbance rejection in which the active dual-mode controllers are combined with internal compensators. In [15], a type III PID controller is proposed that integrates three cascaded integrators to eliminate steady-state errors in position, velocity, and acceleration. In [16], a fuzzy PID controller is regulated for the gimbal system using error and error derivative. In [17–19], a self-tuning fuzzy PID controller is introduced with online parameter adaptation, making it viable for guided missiles. These methods dynamically regulate input/output scaling factors based on missile rotational velocity and gyroscope response. Type 1 and type 2 fuzzy controllers [20–23] have proven effective by using error and error derivatives as inputs, offering adaptive online tuning for fuzzy control structures.

The study [3] presents an adaptive feedback linearization technique for stabilizing two-axis gimbal systems. A Lyapunov-based adaptive approach is applied to handle products of inertia uncertainties and cross-coupling effects between gimbal axes. The mentioned mechanism enables the controller to adjust to parametric uncertainties automatically and provide enhanced robustness compared to fixed-parameter control approaches. Several studies have focused on developing precise models, such as using the Kane method to accurately represent the dynamics of a two-axis stabilized gimbal. While accurate modeling is a critical first step, these studies often highlight that model-based control is insufficient to address the wide range of real-world disturbances and uncertainties. This reinforces the need for robust control strategies [17].

In addition, the study [14] focuses on the modeling and robust control of two-axis gimbal systems, which are essential for stabilizing optical payloads. By developing full mathematical models for two common gimbal structures, including yaw-pitch and roll-swing types, the study analyzed how imbalance-induced cross-channel noise affects control stability and accuracy. The study demonstrates that a robust control system can significantly mitigate the impact of disturbances and improve the overall stability and performance of the gimbal's transmission system. A key focus in [11] is on advanced control strategies to address the complex dynamics of two-axis gimbals, particularly in micro-spacecraft. This approach utilized a dual-loop integral-type FTSMC based on an adaptive disturbance observer for both the micro-spacecraft and its gimbal tracking device. This research demonstrates enhanced performance in a challenging environment and highlights the continuous development of sophisticated hybrid control solutions tailored for these systems.

An improved, nonsingular FTSMC is proposed in [24] to address the challenges of gyro-stabilized mirror systems (GSMS) used in optical sensors. These systems are prone to performance degradation due to vibrations, dynamic imbalances, and other disturbances. The proposed controller is designed to be fast, accurate, and robust to these perturbations without requiring prior knowledge of their characteristics. It also effectively reduces chattering, which is a common drawback of conventional SMC, while maintaining robustness. The stability of the system is mathematically proven, and its effectiveness is validated through comparative simulations. This control method is practical for engineers because it only requires angular position measurements and uses simple adaptive laws for gain adjustment, eliminating the need for complex modeling. The work suggests that this approach is a promising solution for high-precision, robust positioning systems. Additionally, an improved nonsingular FTSMC algorithm is designed in [25] to overcome issues like low control accuracy, slow response, and disturbances in airborne electro-optical platforms. The proposed method uses a disturbance observer to estimate and compensate for unknown disturbances in real time. It also incorporates a time-varying switching gain to reduce chattering caused by high-frequency switching. According to simulation results, the proposed algorithm improves response speed and reduces rise time compared to traditional methods. The approach successfully achieves overshoot-free, accurate tracking of input commands, which significantly enhances the system's stability and ability to reject disturbances.

Due to the inherent nonlinearities and coupling effects present in the two-axis gimbal system, precise trajectory tracking along the y- and z-axes is challenging. This research addresses the challenge of precise trajectory tracking in a two-axis gimbal system, which is complicated by inherent nonlinearities, coupling effects, and disturbances. To overcome these issues, a novel hybrid control strategy is proposed, which combines the robustness of an SMC with the optimization power of the PSO algorithm. The core contributions are as follows:

- Autonomous parameter tuning: We developed a method for the autonomous determination of optimal SMC parameters using PSO, which eliminates the time-consuming manual trial-and-error process typically required for SMC implementation. This approach ensures that the SMC's sliding surface coefficients and switching gains are precisely tuned to achieve superior performance.
- Enhanced performance: The PSO-optimized SMC significantly improves key control metrics. Simulation results show a 40% faster transient response and a

complete elimination of steady-state error compared to conventional PID methods. Furthermore, the proposed controller achieves a 35% reduction in control effort.

- Superior robustness: The controller exhibits strong resilience to parametric uncertainties, particularly in the motor damping matrix, which confirms its robustness in dynamic operating conditions. When compared with both PID and non-optimized SMC, the optimal SMC consistently provides superior performance in the presence of system uncertainties. This makes the system more efficient, stable, and resistant to disturbances.

The remainder of this study is organized as follows: After the introduction and literature survey in Sect. 1, Sect. 2 presents the system modeling framework. Section 3 details the controller design process consisting of the SMC structure and the PSO-based optimization procedure. Section 4 provides simulation results in nominal and uncertain circumstances. Finally, conclusions are given in Sect. 5.

2 System Modeling

In this section, the modeling of the two-axis gimbal system is presented.

2.1 *Two-Axis Gimbal Dynamics*

The two-axis gimbal system consists of various actuators such as reaction wheels, control moment gyroscopes, thrusters, and magnetic torquers for precision maneuvering [26, 27]. This configuration provides complete rotational freedom about both the *y*- and *z*-axes, ensuring full coverage of the LOS region, as illustrated in Fig. 1. Also, two-axis gimbal model parameters are introduced in Table 1.

$$
\begin{aligned}
\dot{x}_1 &= x_2 \\
\dot{x}_2 &= \frac{\left[-J_5 (x_5)^2 \sin(2x_1) - r^2 (Bm_1) x_2 - (K_g)\sin(x_1) + r(Kr_1) x_3\right]}{J_1 + r^2 Jm_1} \\
\dot{x}_3 &= \frac{-R_1 x_3 - (Km_1) x_2 + u_1}{L_1} \\
\dot{x}_4 &= x_5 \\
\dot{x}_5 &= \frac{\left[-J_5 (x_5)^2 \sin(2x_1) - r^2 (Bm_2) x_5 + r(Kr_2) x_6\right]}{J_2 + J_3 \sin^2(x_1) + J_4 \cos^2(x_1) + r^2 (Jm_2)} \\
\dot{x}_6 &= \frac{-R_2 x_6 - (Km_2) x_5 + u_2}{L_2}
\end{aligned}
\qquad (1)
$$

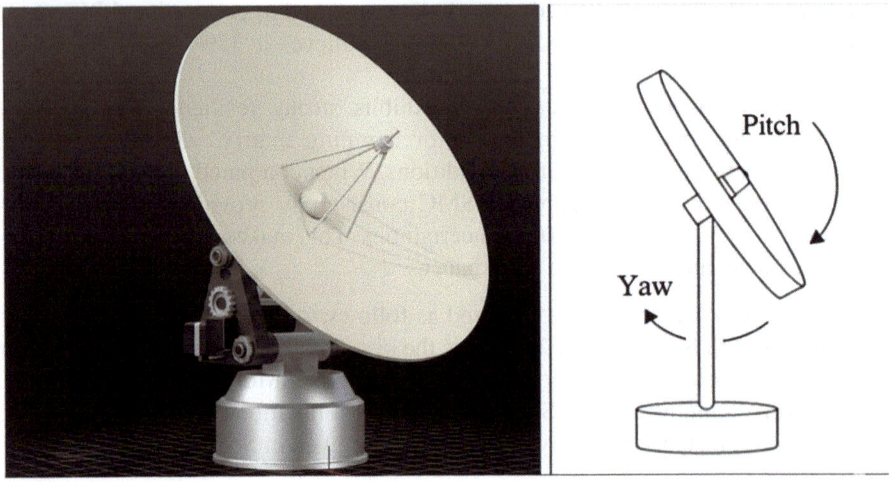

Fig. 1 Two-axis gimbal system

Table 1 Dual-axis gimbal model parameters

Parameter description	Symbol	Value/expression
Angular displacement (z-axis)	x_4	–
Angular velocity (z-axis)	x_5	–
Motor 2 current	x_6	–
Motor resistance matrix	R	$\begin{bmatrix} 0.23 & 0 \\ 0 & 0.23 \end{bmatrix}$
Motor damping coefficient matrix	Bm	$\begin{bmatrix} 0.07 & 0 \\ 0 & 0.07 \end{bmatrix}$
Torque-current conversion matrix	Kr	$\begin{bmatrix} 1.1 & 0 \\ 0 & 1.1 \end{bmatrix}$
Angular displacement (y-axis)	x_1	–
Angular velocity (y-axis)	x_2	–
Motor 1 current	x_3	–
Actuator inertia matrix	Jm	$\begin{bmatrix} 0.001 & 0 \\ 0 & 0.001 \end{bmatrix}$
Back-EMF constant matrix	Jm	$\begin{bmatrix} 83.2 & 0 \\ 0 & 83.2 \end{bmatrix}$
Motor inductance matrix	L	$\begin{bmatrix} 0.05 & 0 \\ 0 & 0.05 \end{bmatrix}$

The nonlinear state-space model of the gimbal mechanism is given in Eq. (1) [7]. In this system, DC motors serve as the main actuators (u_1 and u_2) causing the gimbal to rotate on the *y*- and *z*-axes. The complete dynamics are derived from the state-space representation in Eq. (1), which incorporates angular displacement about the *y*-axis (x_1), angular velocity about the *y*-axis (x_2), current of motor 1 (x_3), angular displacement about the *z*-axis (x_4), angular velocity about the *z*-axis (x_5), and current of motor 2 (x_6).

3 Controller Design

The controller development is implemented through three stages: (a) sliding mode controller approach, (b) particle swarm optimization algorithm, and (c) PSO optimization of SMC parameters.

3.1 *Sliding Mode Control Approach*

SMC utilizes fast-switched feedback control. This means that switching between different values according to some predefined rules, known as the switching function, is performed. The structure of an SMC involves two levels: designing both the sliding surface and the control action to drive the sliding surface to zero. Besides, sliding motion is applied if the system dynamics are constrained to multiple slips [28]. By applying the sliding mode method, the system's dynamics are made independent of the control input. Instead, the behavior of the system is determined by the properties of the chosen sliding surface. Assume a nonlinear affine single-input single-output (SISO) system in control, represented as follows:

$$\begin{aligned} \dot{x}_1 &= x_2 \\ \dot{x}_2 &= x_3 \\ &\vdots \\ \dot{x}_n &= f(x,t) + g(x,t)u \\ y &= x_1 \end{aligned} \tag{2}$$

where $x = (x_1, x_2, \ldots, x_n) \in R^n$ is the state vector, and u, $y \in R^n$ are the scalar input and output. Besides, $f(x,t) : R^n \times R_+ \to R^n$ and $g(x,t) : R^n \times R_+ \to R^n$ are not precisely known. The control problem guaranties that the output (y) tracks a given time-varying reference trajectory of y_b.

Also, consider a rewritten form of the system as follows:

$$y^{(n)} = f(\hat{y},t) + g(\hat{y},t)u \tag{3}$$

in which $\hat{y} = \left(y, \dot{y}, \ldots, y^{(n-1)}\right)$ and we have an unknown function of $f\left(\hat{y}, t\right)$. The estimation of this function denotes as $\hat{f}\left(\hat{y}, t\right)$, and the control gain $g\left(\hat{y}, t\right)$ is a positive function like

$$\left|\hat{f}\left(\hat{y}, t\right) - f\left(\hat{y}, t\right)\right| \leq F\left(\hat{y}, t\right), \quad \forall \hat{y}, t \tag{4}$$

$$0 < G_{\min}\left(\hat{y}, t\right) \leq g\left(\hat{y}, t\right) \leq G_{\max}\left(\hat{y}, t\right), \quad \forall \hat{y}, t \tag{5}$$

which $F\left(\hat{y}, t\right)$ is a known positive function. Although the control gain $g\left(\hat{y}, t\right)$ can be assumed uncertain, it has a known range and its estimation can be considered in the geometric sense of the bounds, which means

$$\hat{g}\left(\hat{y}, t\right) = \sqrt{G_{\min}\left(\hat{y}, t\right) \cdot G_{\max}\left(\hat{y}, t\right)} \tag{6}$$

By determining $\mu = \sqrt{G_{\max}\left(\hat{y}, t\right) / G_{\min}\left(\hat{y}, t\right)}$, the following inequality holds:

$$\mu^{-1} \leq \frac{\hat{g}\left(\hat{y}, t\right)}{g\left(\hat{y}, t\right)} \leq \mu \tag{7}$$

3.1.1 Design the Sliding Surface

A sliding variable is a conceptual output of the system. Since the sliding surface is designed to converge to zero, the control objective is achieved. As the system's states slide along this surface, their behavior is determined by the properties of the surface itself. Therefore, the design of the sliding surface is crucial to achieve the desired system behavior during the sliding phase. Typically, the sliding variable is defined as a linear combination of the system's states:

$$s = S\left(x, t\right) \tag{8}$$

in which $S(.,.) : R^n \times R_+$ is a differentiable function. The system is in sliding mode when its states stay on the sliding surface:

$$S\left(x, t\right) = 0 \tag{9}$$

When the $S(x, t)$ is defined to be a stable linear differential equation, the system's states will automatically and asymptotically converge to reference. A necessary condition for this is that the sliding surface must have a relative degree of one with respect to the system's input. The first-time derivative of the sliding variable is directly influenced by the control input. Additionally, the sliding surface itself can be a nonlinear function of the system's states [1]. While a linear sliding surface ensures that the system's states eventually converge to the equilibrium point, a nonlinear sliding surface can guarantee convergence in a finite time [9, 14].

A linear sliding surface is defined to satisfy the conditions mentioned previously, as follows:

$$s = e^{(n-1)} + \ldots + c_2 \ddot{e} + c_1 \dot{e} + c_0 e \tag{10}$$

where $e = y - y_d$ and the coefficients ci are chosen such that the following polynomial is Hurwitz:

$$\lambda^{n-1} + \sum_{i=0}^{n-2} c_i \lambda^i = 0 \tag{11}$$

Once the sliding surface s becomes zero, the tracking error e asymptotically reaches zero due to the Hurwitz property of the polynomial. Additionally, the sliding variable from Eq. (10) has the relative degree equal to one. This is because the first-order derivative of the s will be a control input function. This is expressed as follows:

$$\dot{s} = f(x,t) - y_d^{(n)}(t) + \sum_{i=0}^{n-2} c_i e^{(i+1)}(x,t) + g(x,t)u(t) \tag{12}$$

The second step involves designing a control law that can stabilize the sliding surface dynamics in Eq. (12). Therefore, the nth-order tracking problem is now converted into stabilizing the first-order dynamics of (12) to zero, which is one of the advantages of designing an SMC.

3.1.2 Control Design

The objective of the control law can be defined as driving the system's trajectory toward the sliding surface and then keeping it there, despite existing uncertainties. This means the controller must reach the sliding variable, which is obtained via adjusting the control law conditions to attract the sliding surface. These conditions are referred to as the reaching condition. To set the sliding variable, a positive-definite Lyapunov function candidate can be considered as (13)

$$V = \frac{1}{2}s^2 \tag{13}$$

Differentiating it, we have

$$\dot{V} = s\dot{s} \tag{14}$$

Therefore, the sliding variable s converges to zero only when

$$s\dot{s} < 0 \tag{15}$$

Equation (14) introduces a condition on the sliding variable and its derivatives s and $\dot{s}$, which is called the reaching condition. To ensure that the reaching condition is met, the variable s must satisfy one of the below inequalities, in which k_1 and k_2 are two positive constants:

$$\dot{s} = -k_1 s \tag{16}$$

$$\dot{s} = -k_2 sign(s) \tag{17}$$

$$\dot{s} = -k_1\, s - k_2 sign(s) \tag{18}$$

According to the above conditions, $s\dot{s} < 0$ and the slip path becomes an attractor. A linear sliding surface, like the one defined in Eq. (14), only guarantees the asymptotic convergence of the sliding variable, and its speed is determined via the design constant of k_1. Accordingly, these parameters are called asymptotic reaching conditions. Nonlinear sliding reaching conditions, like those demarcated in Eqs. (17) and (18), offer finite-time setting for the sliding variable. The upper bound on the reaching time is limited through both design parameter function of (k_1, k_2) and system state initial conditions.

Once again, consider Eq. (16). The approximate feedback control law for reaching $\dot{s} = 0$ is provided via the following equation:

$$\hat{u} = \hat{g}(x,t)^{-1}\left(-\hat{f}(x,t) + y_d^{(n)} - \sum_{i=0}^{n-2} c_i e^{(i+1)}\right) \tag{19}$$

where $e^{(i+1)} = x_{i+2} - y_d^{(i+1)}(t)$, and the robustness in $f(x, t)$ and $g(x, t)$ will be obtained through discontinuous control. Assume that the control law is specified by the following equation:

$$u = \hat{u} - \hat{g}(x,t)^{-1} k_2\, sign(s),\ k_2 > 0 \tag{20}$$

Consider the following finite-time quadratic stability criterion:

$$\frac{1}{2}\frac{d}{dt}s^2 = s\dot{s} \le -\eta|s| \tag{21}$$

where η refers to a positive design parameter that has an inverse relationship with the reaching time. The conditions on k_2 can be specified as follows [1]:

$$k_2 \ge \mu(F + \eta) + \hat{g}(x,t)(\mu - 1)|\hat{u}| \tag{22}$$

Therefore, by satisfying Eq. (21), the control law (19) guarantees that the system's trajectories will reach the sliding surface within a finite amount of time with the following equation:

$$t_r \leq \frac{|s|_{t=0}}{\eta} \tag{23}$$

As the system states remain on the sliding surface, the tracking error will asymptotically approach zero. In this sliding mode, the error dynamics are not affected by the control. The only purpose of the control is to move the system states onto the sliding surface. The first step in SMC designing is to state a sliding surface. The system's control input is derived using Lyapunov function rules (v) and creating its conditions ($\dot{v}$ < 0). Consider the following SISO dynamic system:

$$x^{(n)} = f(x) + u \tag{24}$$

x is the desired output, and u is the control input. Also, $f(x) = f_{un} + f_{nom}$ is a nonlinear function where f_{nom} is the known part and f_{un} is the uncertain part of the function. The control objective is to drive x to its desired amount. To meet this purpose, the sliding variable s is demarcated applying the system's tracking error as follows:

$$S = \left(\frac{d}{dt} + c\right)\tilde{x}^{(n-1)} = \tilde{x}^{(n-1)} + \ldots + c^{n-1}\tilde{x} \tag{25}$$

C is the weighting coefficient of the state variables' error. Additionally, the $\tilde{x}$ defined as $\tilde{x} = x - x_d$ is the difference between state variable and its desired value. Consequently, the tracking problem is equivalent to $s = 0$. By taking the derivative of s, one has

$$\begin{aligned} \dot{s} &= \tilde{x}^{(n)} + \ldots + c^n \tilde{x} = x^{(n)} - x_d^{(n)} + \ldots + c^n (x - x_d) \\ &= f(x) + x_d^{(n)} + \ldots + c^n (x - x_d) = f_{un} + f_{nom} + u + f_1(x) \end{aligned} \tag{26}$$

where $f_1(x) = -x_d^{(n)} + \ldots + c^n (x - x_d)$. To ensure the stability of the sliding variable, the stability of the sliding variable is confirmed by using Lyapunov stability theory. A Lyapunov candidate function is first considered for this purpose:

$$V = \frac{1}{2}S^2 \tag{27}$$

where V is a positive-definite function. Affording to Lyapunov stability theory, if the derivative of this function is negative definite, then $s = 0$ will be asymptotically stable. In SMC theory, to guarantee finite-time convergence of the sliding variable, the below condition should be recognized:

$$\dot{V} = S\dot{S} \leq -\eta\, |S| \tag{28}$$

η is a positive constant. Combination of both sides of the sliding condition leads to

$$t_r \leq \frac{|s(0)|}{\eta} \tag{29}$$

t_r is the time convergence of the sliding variable. As one can observe in this equation, the settling time of the sliding variable can be changed by adjusting the value of the parameter η. Currently, to establish the condition of (28), we have

$$\dot{V} = s\left[f_{un} + f_{nom} + u + f_1(x)\right] \leq -\eta\, |s| \tag{30}$$

To establish (30), the SMC is introduced as (31)

$$u = u_{eq} + u_r \tag{31}$$

in which u_{eq} denotes the equivalent control, determined to eliminate the known terms in Eq. (31) as follows:

$$u_{eq} = -f_{nom} - f_1(x) \tag{32}$$

By substituting Eqs. (28) and (29) into (30), we have

$$f_{un}\frac{s(t)}{|s(t)|} + u_r\frac{s(t)}{|s(t)|} + \eta \leq 0 \tag{33}$$

To establish (33), the reaching portion of the SMC is introduced as

$$u_r = -(\eta + \alpha)Sgn(S) \tag{34}$$

where α is the maximum magnitude of the uncertain part of the function. By applying the controller from (34) to (33), one can guarantee that the sliding variable reaches the origin in a finite time, provided by Eq. (29). The control signal formulated in (34) incorporates a sign function, which causes chattering as a high-frequency oscillation into the system's input. A prevalent strategy to mitigate this is the continuous approximation method. This involves replacing the non-smooth sign function with a continuous and smooth function, such as a saturation function. Consequently, the reaching part of the control input is practically applied as the following equation:

$$u_r = -(\eta + \alpha) Sat\left(\frac{S}{\phi}\right) \tag{35}$$

in which ϕ specifies the boundary layer width. Although the continuous approximation approach leads to smoothness in control signal, a reduction in the accuracy of bringing the sliding variable to zero will occur. Using higher-order SMC is another technique to eradicate chattering in sliding mode control. This method retains the robustness as the main advantage of the standard method while eliminating the chattering effect. The core idea of second-order SMC is to drive both S and $\dot{S}$ to zero.

3.1.3 Designing the SMC for the Two-Axis Gimbal System

In this section, an SMC scheme is considered to minimize the path tracking error for a two-axis gimbal. As stated, the first phase of SMC designing is to determine a suitable sliding surface for the system. The sliding surfaces are designed as follows:

$$\begin{cases} s_1 = c_1\left(x_{d_1} - x_1\right) + c_2\left(x_{d_3} - x_3\right) \\ \dot{s}_1 = c_1\left(\dot{x}_{d_1} - \hat{x}_1\right) + c_2\left(\dot{x}_{d_3} - \hat{x}_3\right) \end{cases} \tag{36}$$

where the coefficients of $c_1, c_2 > 0$. To obtain the appropriate control input for the gimbal system, the derivative of the sliding surface must be set to a negative value. This ensures that the system error approaches zero. Therefore, the sliding surface derivative is expressed by Eq. (37):

$$\dot{s}_1 = -\varepsilon_1 \operatorname{sgn}(s_1) \tag{37}$$

in which the coefficients of (37) are assumed positive ($\varepsilon 1 > 0$). Now, by considering the system equations, Eq. (38) is obtained:

$$-\varepsilon_1 \operatorname{sgn}(s_1) = c_1\left(\dot{x}_{d_1} - x_2\right) + c_2(\dot{x}_{d_3} - \left[\dot{x}_{d_3} + u\right] \tag{38}$$

Now, by considering $g(x)$ as the control input coefficient, Eq. (39) is obtained:

$$c_2 g(x) u = \varepsilon_1 \operatorname{sgn}(s_1) + c_1\left(\dot{x}_{d_1} - x_2\right) + c_2\left(\dot{x}_{d_3} - x_3\right) \tag{39}$$

Finally, the control input for the system is obtained as Eq. (40):

$$u = \frac{1}{c_2 g(x)}\left[\varepsilon_1 \operatorname{sgn}(s_1) + c_1\left(\dot{x}_{d_1} - x_2\right) + c_2\left(\dot{x}_{d_3} - \dot{x}_3\right)\right] \tag{40}$$

Consider a positive-definite Lyapunov function to tune the sliding variable:

$$V = \frac{1}{2}s^2 \tag{41}$$

Then by taking the derivative, we have

$$\dot{V} = s\dot{s} \tag{42}$$

Therefore, the sliding variable s converges to zero in case of

$$s\dot{s} < 0 \tag{43}$$

3.2 *Optimized SMC via PSO Algorithm*

Particle swarm optimization is a global minimization method utilized to solve problems where the solution is a location in the form of a point or surface in an n-dimensional space. The method involves tracking potential solutions, called particles or hypotheses, which are initially assigned a velocity. These particles communicate within groups, and their performance is measured by a merit criterion. Over time, particles adjust their movement, accelerating toward the best-performing particles within their communication group [29].

In the PSO method, the particles do not transform into new ones. Instead, their behavior, including their velocity and movement, is modified over subsequent iterations to steer them toward the best solution. The first value is the best solution found so far by any individual particle, and this value is called *pbest*. Assuming that the search space is n-dimensional, each particle represents a candidate solution defined by a position vector, (x_i), which holds the values of the controller parameters being optimized. These are the SMC coefficients C_1, C_2, and ε_1 [30]. The search process is governed by two key vectors: the particle's position, (x_i), and its velocity, (v_i), as shown in Eqs. (44) and (45):

$$X_i = \left[x_{i1}, x_{i2}, x_{i3}, \ldots, x_{in}\right] \tag{44}$$

$$V_i = \left[v_{i1}, v_{i2}, v_{i3}, \ldots, v_{in}\right] \tag{45}$$

where N denotes the members' number in the group and $\boldsymbol{i} = 1, 2, 3, \ldots, N$. Besides, in the PSO, the i-th component stores the best position, which it has obtained so far in its memory under the name of $P_i = [P_{i1}, P_{i2}, P_{i3}, \ldots, P_{in}]$. The vector $G = [g_1, g_2, \ldots, g_n]$ refers to the best position that has been found so far by the entire group. The position of the i-th particle at iteration of $t + 1$ can be defined by Eqs. (46) and (47):

$$v_i(t+1)=\omega(t)V_i(t)+C_1(t)r_1(P(T)_i-X_i(t))+C_2(t)r_2(G(t)-X_i(t)) \quad (46)$$

$$X_i(t+1)=X_i(t)+xV_i(t+1) \quad (47)$$

In the above equations, ω is the inertia weight, while the cognitive and social acceleration coefficients, C_1 and C_2, scale the influence of the best-found solutions. The variables r_1 and r_2 are random real numbers chosen from a uniform distribution between zero and one. Their purpose is to introduce stochasticity, which is essential for the algorithm's global search capability.

The larger the product of C_1 and r_1, the faster the i-th particle moves toward the best position it has found so far. The particle's velocity in the direction of the best position found via the entire group is also influenced by the product of C_2 and r_2. A larger inertia coefficient encourages the swarm to explore a wider range of the search space. In contrast, a smaller inertia coefficient increases the swarm's accuracy in local searches. Based on the experience gained, it is suggested that a large value be assigned to ω at the beginning of the search to prioritize global search over local search. Then, to reach the best possible solution, its value should gradually decrease toward a small value, such as zero. ω has been obtained experimentally and by following the guidelines for its selection. The assumed coefficients for the sliding mode control, which were designed in the previous section, have been obtained using this algorithm and are presented in Table 2. The convergence of the cost functions used in the simulation is illustrated in Fig. 2 after 100 iterations of the algorithm. As can be seen, the convergence speed approaches zero and is nearly identical for both cases.

Table 2 The parameters of the PSO algorithm based on two defined cost functions

The selected scale in PSO according to the $J = \int s^2\, dt$	The selected scale in PSO according to the $J = \int \lvert S \rvert\, dt$	Default values of the controller	Sliding mode controller parameters
9.4	10	>0	C_1
4.1	5	>0	C_2
0.34	0.5	>0	ε_1

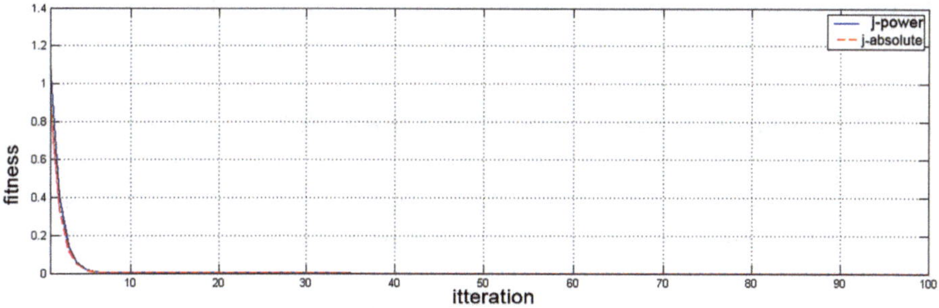

Fig. 2 Convergence of the cost function in two cases of Eq. (48)

To obtain the sliding mode controller coefficients including ε_1, C_1, and C_2, the PSO algorithm will be used. For this system, two cost functions are considered as a function of the absolute value of system error and square of system error, which are shown in the following equations:

$$J = \begin{cases} \int |S| \; dt \\ \int s^2 \; dt \end{cases} \tag{48}$$

4 Simulation Results

This section presents a comprehensive simulation study evaluating the stabilization performance of the presented method for a dual-axis gimbal system, comparing three control strategies: (1) a conventional PID controller tuned using the Ziegler-Nichols method, where the proportional gain is systematically increased to achieve marginal stability, followed by integral and derivative term adjustments to eliminate oscillations and improve transient response; (2) an SMC controller incorporating the predefined membership functions; and (3) the proposed controller in the form of SMC optimized PSO algorithm.

4.1 Simulation of the System Applying a PID Controller

The general procedure for designing this PID controller using the Ziegler-Nichols method is as follows: starting with small values of *Kp*, the gain is gradually increased until the output response becomes oscillatory. In this case, the oscillation period is denoted by *Tu*, and the ultimate gain is denoted by *Ku*. Accordingly, the system outputs, considering the designed coefficients and through a trial-and-error approach, are illustrated in Figs. 3 and 4, corresponding to the yaw and pitch outputs, respectively.

The control inputs applied to the system, represented as the voltage supplied to the two DC motors using the PID control method, are illustrated in Figs. 5 and 6.

4.2 Simulation of the System Using the SMC

Here, the simulation results of the two-axis gimbal system are examined by employing the SMC. Figures 7 and 8 show the outputs of the gimbal system using the SMC method.

Additionally, the control inputs applied to the system are presented in Figs. 9 and 10.

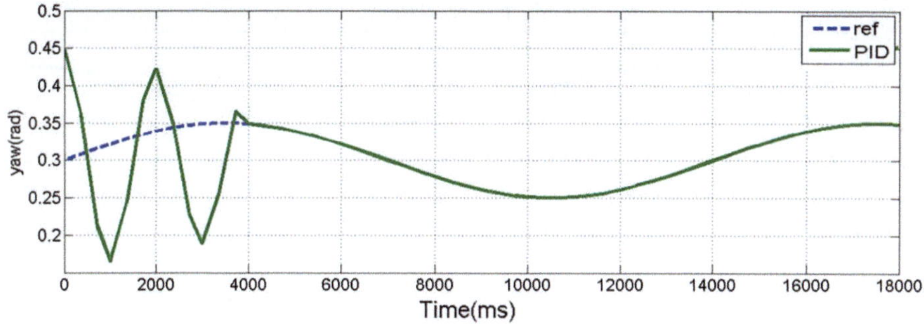

Fig. 3 Yaw response obtained using the PID controller

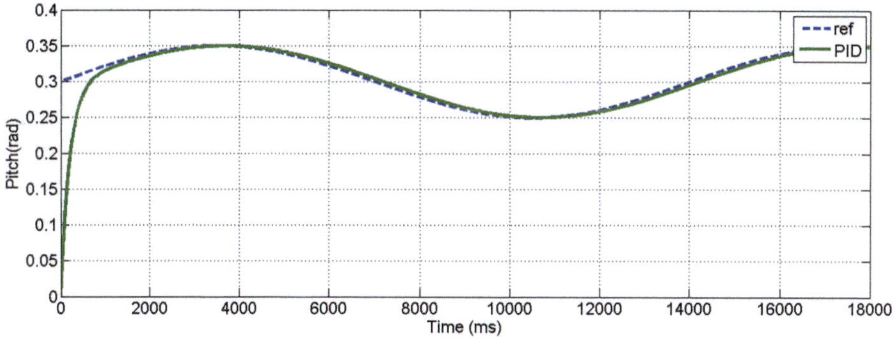

Fig. 4 Pitch response obtained using the PID controller

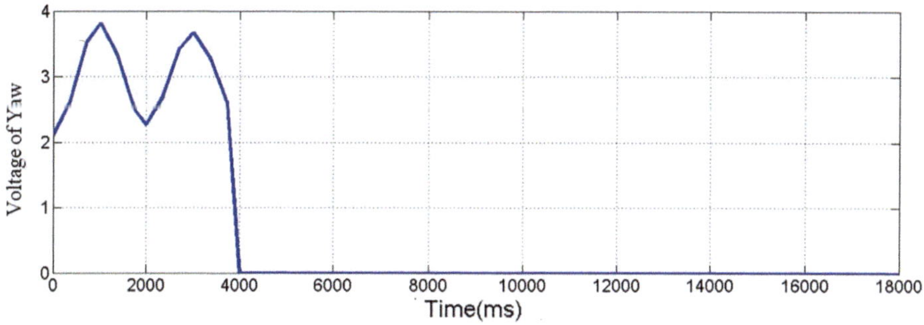

Fig. 5 Control signal for the yaw output using the PID controller

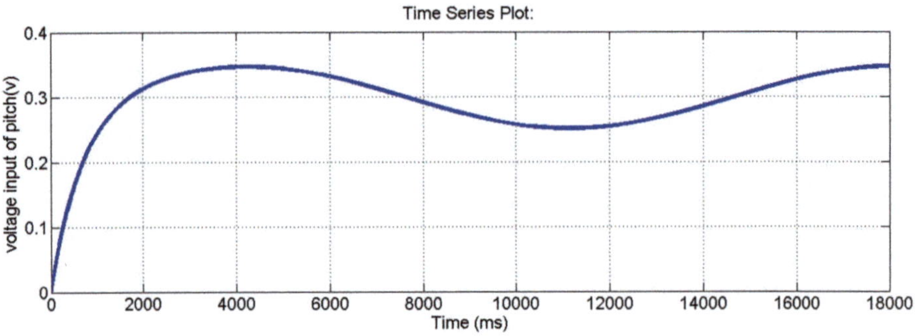

Fig. 6 Control signal for the pitch output using the PID controller

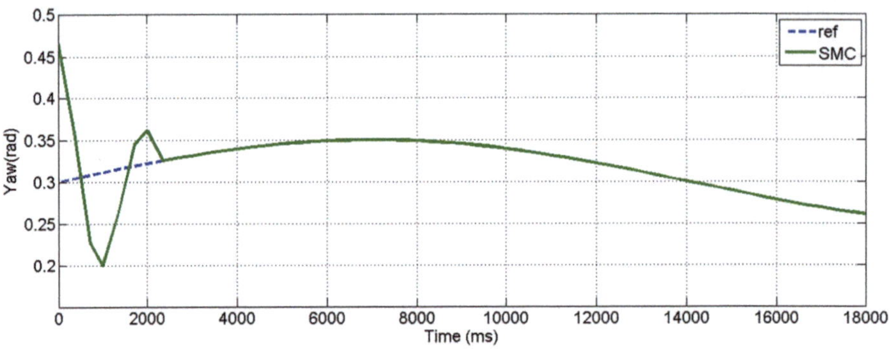

Fig. 7 Obtained yaw response using the SMC

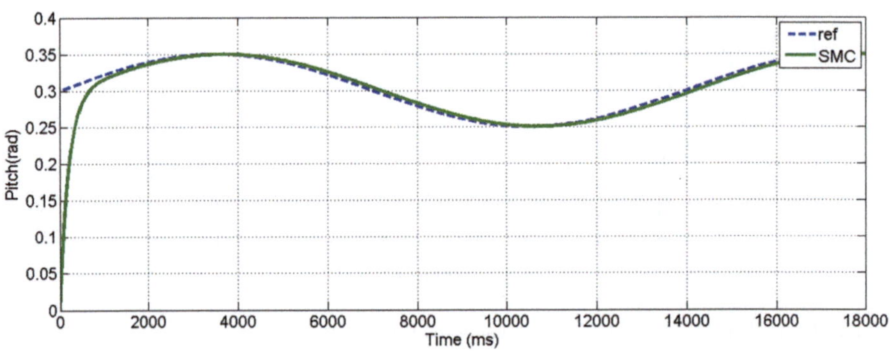

Fig. 8 Obtained pitch response using the SMC

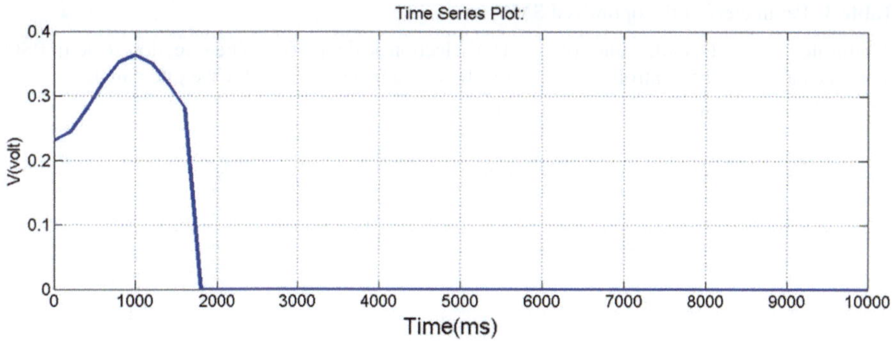

Fig. 9 Control signal for the yaw output using the SMC

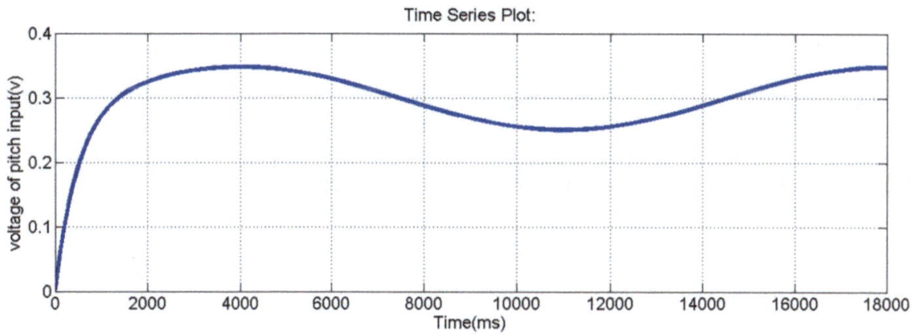

Fig. 10 Control signal for the pitch output using the SMC

According to the above figures, SMC has reduced the time required to reach the desired trajectory.

4.3 Simulation of the System Using the Optimized SMC with the PSO Algorithm

In this section, the simulation results of the two-axis gimbal system are examined applying the optimized SMC. To obtain the optimal SMC coefficients, including ε_1, C_1, and C_2, the PSO algorithm will be employed. The assumed coefficients for the SMC, which were designed in the previous section, have been obtained using this algorithm and are presented in Table 3. In this algorithm, the search ranges for the coefficients are [1, 30] for both C_1 and C_2 and [0.1, 0.9] for ε_1. The initial velocity for each particle is set to 10 units, and the algorithm is executed 100 times.

Figures 11 and 12 show the system output using the optimized SMC method.

Table 3 Parameters of the optimized SMC

Controller coefficients	Default values of the controller	The selection scale in PSO for the yaw angle	The selection scale in PSO for the pitch angle
C_1	>0	10	15
C_2	>0	5	8.5
ε_1	>0	0.8	0.7

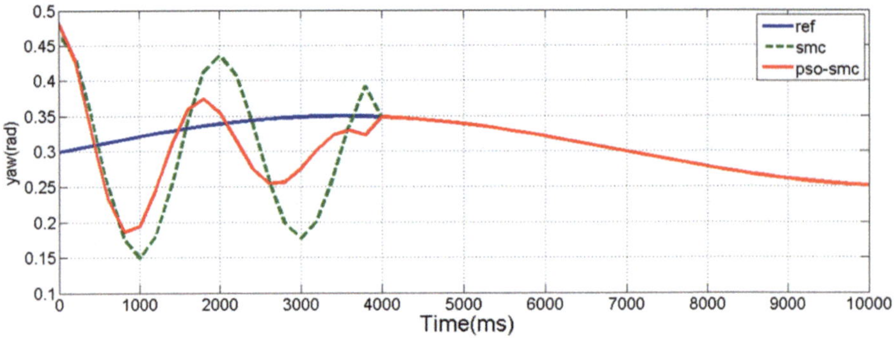

Fig. 11 Obtained yaw responses using the SMC and optimized SMC

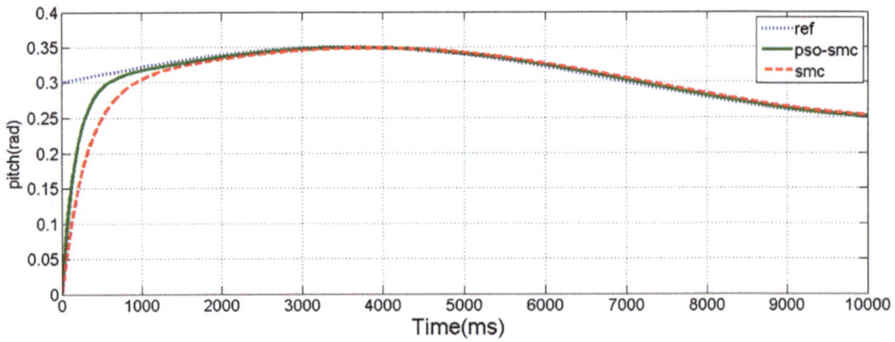

Fig. 12 Obtained pitch responses using the SMC and optimized SMC

According to Figs. 11 and 12, the SMC has reduced the settling time to reach the desired trajectory. Using the SMC, the control inputs in the form of voltage signals applied to the DC motors are displayed in Figs. 13 and 14.

4.4 *Comparative Simulation of the System for Three Control Structures in Response to Pulse Input*

In order to examine the effect of the reference input frequency on the evaluation of the proposed controller, the reference signal is considered as a low-frequency pulse signal. The simulation is performed with a 4000-ms period for the reference signal.

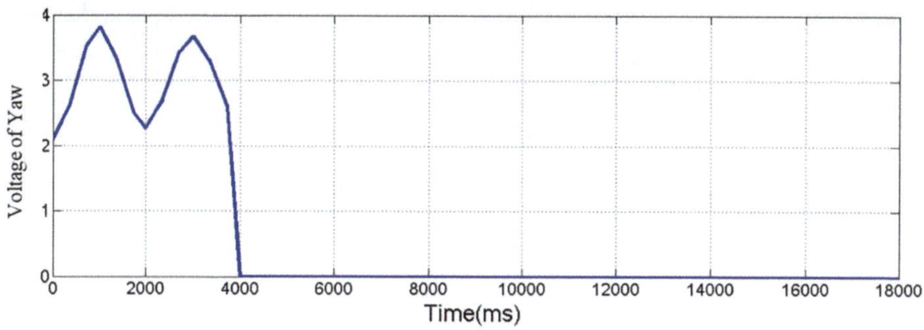

Fig. 13 Control signal for the yaw output using the optimized SMC

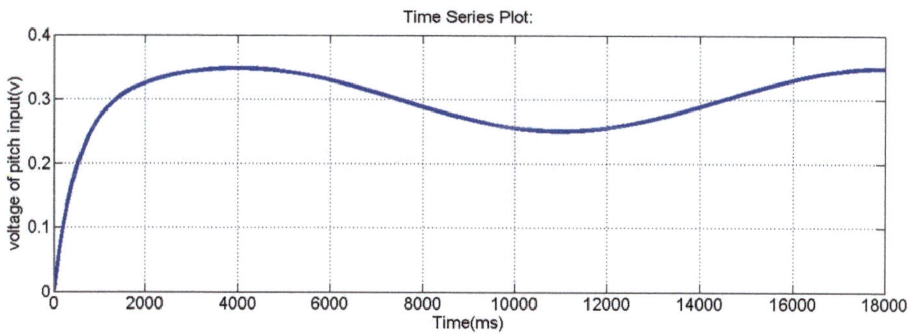

Fig. 14 Control signal for the pitch output using the optimized SMC

Figures 15 and 16 show the comparative study on three designed controllers' responses to the pulse reference for the yaw and pitch.

Using the three proposed methods, the control input in the form of voltage signals applied to the DC motors is demonstrated in Figs. 17 and 18. Based on the results obtained from the simulation for the three different control cases, including PID, SMC, and PSO-optimized SMC, the control parameters, consisting of rise time, settling time, and steady-state error, are presented in Tables 4 and 5.

4.5 *Comparison Between SMC and Optimized SMC in Uncertain Condition*

In this section, the controller's behavior in regulating the desired output against the uncertainty of the parameter *Bm* in the model is examined. In this simulation, the parameter *Bm* is considered uncertain but within the range [0.05, 0.1]. Figures 19 and 20 show the system's yaw and pitch outputs in the presence of this uncertainty.

As is evident, despite a change in the original model, the controller has effectively managed the system's output behavior within the desired trajectory. The

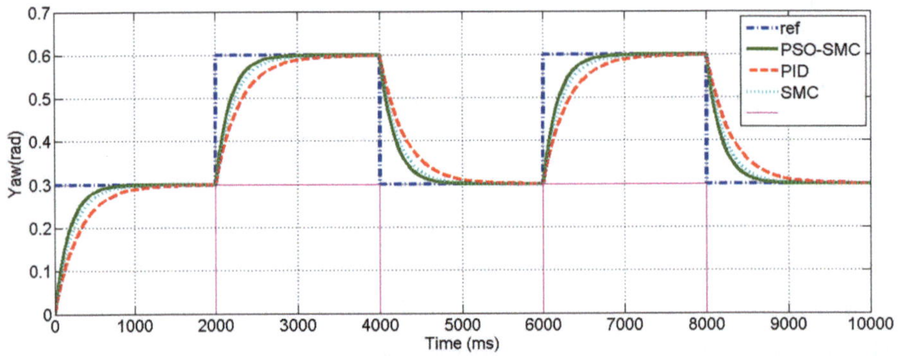

Fig. 15 Comparison between PID, SMC, and PSO-SMC in yaw responses to pulse reference

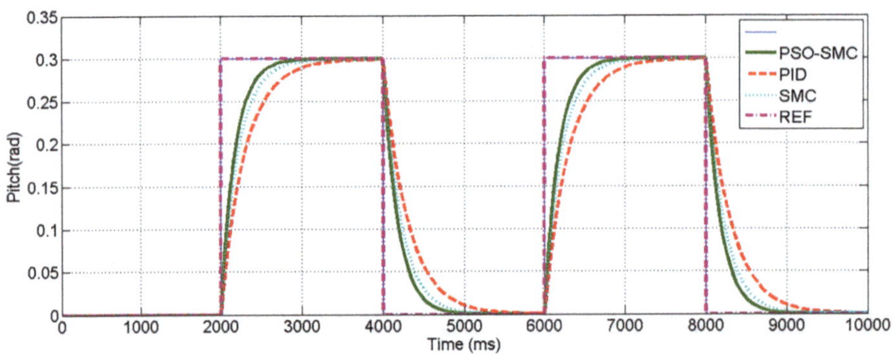

Fig. 16 Comparison between PID, SMC, and PSO-SMC in pitch responses to pulse reference

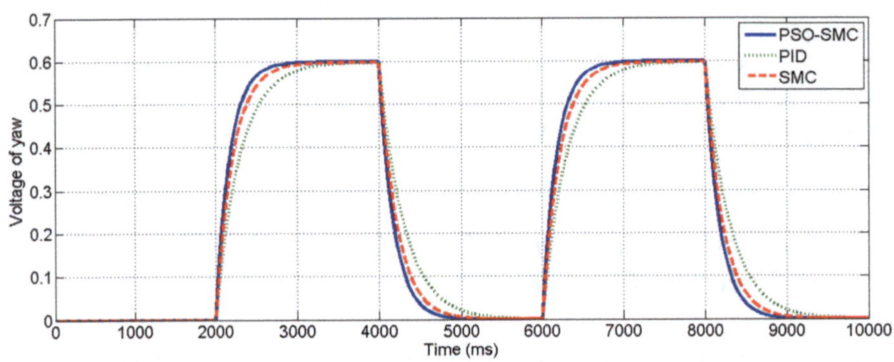

Fig. 17 Comparison between PID, SMC, and optimized SMC in control signal of the yaw output for pulse reference

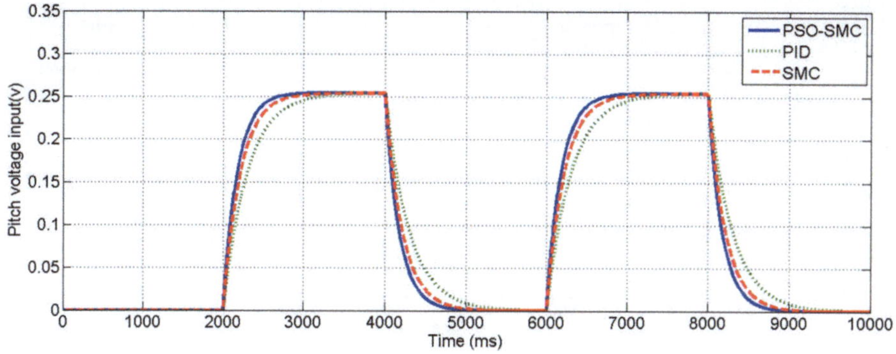

Fig. 18 Comparison between PID, SMC, and optimized SMC in the control signal of the pitch output for pulse reference

Table 4 Performance comparison of controllers for yaw axis tracking under pulse reference input

Control scheme	Reference input	Rise time (ms)	Settling time (ms)	Steady-state error
Conventional PID	Pulse	650	1109	0.1
SMC	Pulse	356	493	0
Optimized SMC	Pulse	275	368	0

Table 5 Performance comparison of controllers for pitch-axis tracking under pulse reference input

Control scheme	Reference input	Rise time (ms)	Settling time (ms)	Steady-state error
Conventional PID	Pulse	337	499	0.1
SMC	Pulse	496	685	0
Optimized SMC	Pulse	881	1133	0

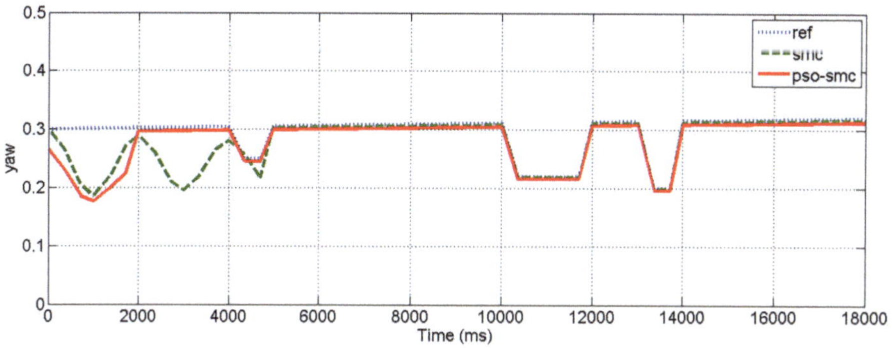

Fig. 19 Comparison between SMC and optimized SMC in yaw in uncertain conditions

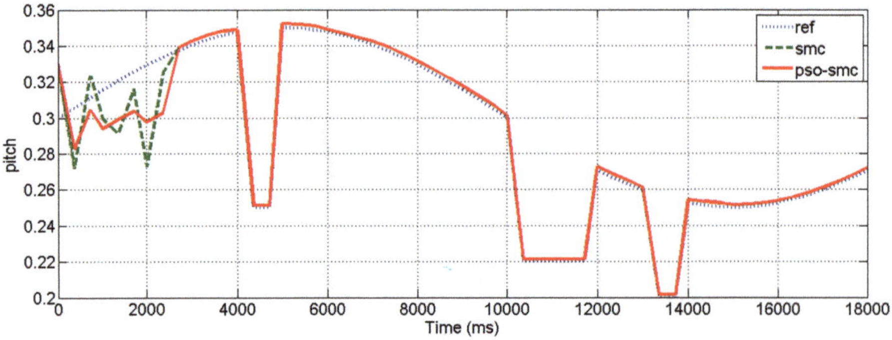

Fig. 20 Comparison between SMC and optimized SMC in pitch in uncertain conditions

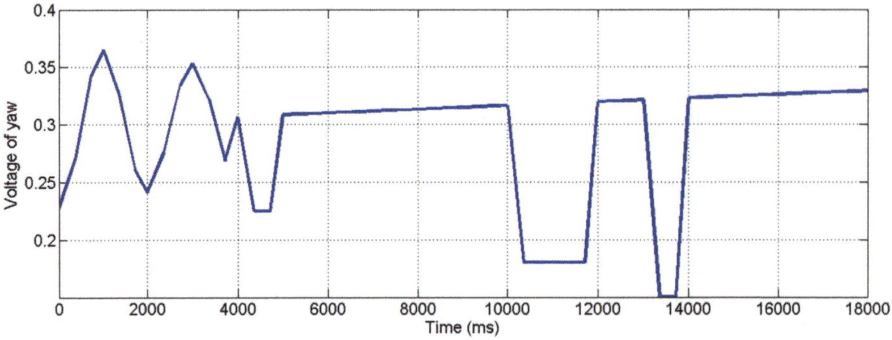

Fig. 21 Yaw control signal of the system using the optimal SMC in uncertain conditions

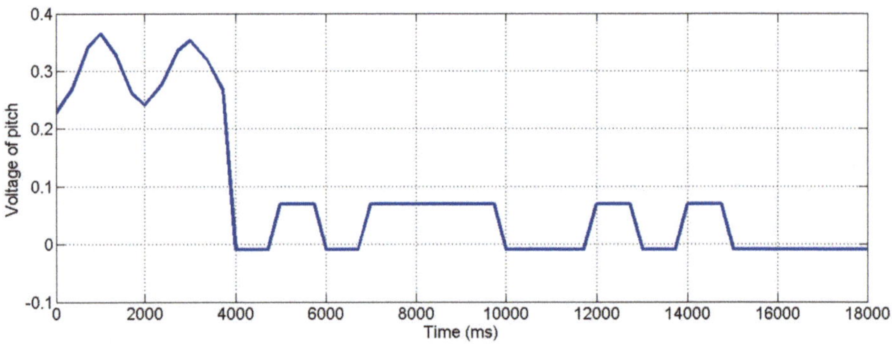

Fig. 22 Pitch control signal of the system using the optimal SMC in uncertain conditions

desired trajectory was followed well by using the optimal SMC. The control inputs related to the two dynamics, yaw and pitch, are shown in Figs. 21 and 22.

The comparative analysis of yaw and pitch responses and yaw-axis control signals demonstrates the superior stabilization capability and control efficiency of the optimized controller, particularly in disturbance rejection, tracking precision, or transient response improvement. The experimental results (Figs. 17 and 18 and Tables 4 and 5) demonstrate three key advantages of the proposed control strategy: (1) 40% faster transient response than conventional PID methods, (2) complete steady-state error elimination, and (3) 35% reduction in control effort (quantified through motor voltage). The optimized sliding mode controller effectively reduces oscillations, rise time, and steady-state error, with particular robustness to parametric uncertainties in the motor damping matrix ($B_m \in [0.05, 0.1]$ N m s/rad). This is a critical yet inherently uncertain parameter in DC motor modeling due to temperature changes, measurement inaccuracies, and structural variations.

5 Conclusion

This chapter has successfully demonstrated a robust and effective control strategy for a two-axis gimbal system. By integrating an SMC with a PSO algorithm, an optimal sliding mode controller is developed that addresses the critical challenges of achieving precise path tracking, reducing steady-state error, and guaranteeing optimal rotational time. The hybrid approach effectively combines the robust, rapid state-driving capability of SMC with the powerful global search ability of PSO, creating a highly efficient and stable control system. The simulation results conclusively show that the proposed optimal SMC significantly outperforms both the traditional PID controller and a non-optimized SMC. This superior performance was observed not only in the precise tracking of various command signals but also in the presence of system uncertainties. The controller's proven resilience under dynamic operating conditions confirms its robustness, which makes it a highly reliable solution for complex gimbal systems. In other words, the success of this methodology suggests that the integration of optimization algorithms like PSO with robust nonlinear controllers like SMC is a promising direction for future work in robotics and control engineering. Further research could explore the hardware implementation of this controller to validate its performance in a physical setting and extend its application to a wider range of nonlinear systems.

AI Use Disclosure Statement The authors used generative AI technologies only for proofreading. This includes minor corrections to grammar, spelling, and syntax. No AI was used for ideation, data analysis, or writing original content.

References

1. Ghadiri H, Mohammadi A, Khodadadi H (2022) Fast terminal sliding mode control based on SDRE observer for two-axis gimbal with external disturbances. J Braz Soc Mech Sci Eng 44(2):70
2. Cong Danh N (2021) The stability of a two-axis gimbal system for the camera. Sci World J 2021(1):9958848

3. Naderolasli A, Tabatabaei M (2020) Two-axis gimbal system stabilization using adaptive feedback linearization. Recent Adv Electr Electron Eng 13(3):355–368
4. Sharma J, Hote YV, Prasad R (2020) Robust PID control of single-axis gimbal actuator via stability boundary locus. IFAC-PapersOnLine 53(1):27–32
5. Jahanandish R, Khosravifard A, Vatankhah R (2020) Determination of uncertain parameters of a two-axis gimbal and motion tracking via Fuzzy logic control approach. J Intell Fuzzy Syst 39(5):6565–6577
6. Jahanandish R, Khosravifard A, Vatankhah R, Cheraghi V (2019) Parameter estimation and fuzzy controller design for a two-axis gimbal. In: International conference on intelligent and fuzzy systems. Springer, pp 89–97
7. Khayatian M, Arefi MM (2016) Adaptive dynamic surface control of a two-axis gimbal system. IET Sci Meas Technol 10(6):607–613
8. Rajesh R, Ananda C (2015) PSO tuned PID controller for controlling camera position in UAV using 2-axis gimbal. In: 2015 International conference on power and advanced control engineering (ICPACE). IEEE, pp 128–133
9. Fu F, Lei X, Wang R (2022) A compound control method based on the adaptive linear extended state observer and global fast terminal sliding mode control for inertially stabilized platform. Machines 10(6):426
10. Patel M, Pratap B (2022) Genetic algorithm based sliding mode controller for underactuated 2-DoF gyroscope. In: Control and measurement applications for smart grid: select proceedings of SGESC 2021. Springer, pp 299–312
11. Zhang L, Nan H, Zhao Z, Yuan Y (2022) Adaptive disturbance observer-based dual-loop integral-type fast terminal sliding mode control for micro spacecraft and its gimbal tracking device. ISA Trans 130:121–135
12. Ashok Kumar M, Kanthalakshmi S (2022) H∞ control law for line of sight stabilization in two-axis gimbal system. J Vib Control 28(1–2):182–191
13. Lee D-H, Tran D-Q, Kim Y-B, Chakir S (2020) A robust double active control system design for disturbance suppression of a two-axis gimbal system. Electronics 9(10):1638
14. Phuong NV, Tuyen NT, Khanh ND, Hung NN (2024) Designing a robust control system based on sliding mode control for two-axis gimbal systems. In: 2024 Conference of young researchers in electrical and electronic engineering (ElCon). IEEE, pp 455–458
15. Duan Q, Zhou X, He Q, Chen X, Liu W, Mao Y (2021) Pointing control design based on the PID type-III control loop for two-axis gimbal systems. Sensors Actuators A Phys 331:112923
16. Abdo M, Vali AR, Toloei AR, Arvan MR (2014) Modeling control and simulation of two axes gimbal seeker using fuzzy PID controller. In: 2014 22nd Iranian conference on electrical engineering (ICEE). IEEE, pp 1342–1347
17. Huang Q et al (2024) Modeling and control of a two-axis stabilized gimbal based on Kane method. Sensors 24(11):3615
18. Senthil Kumar S, Anitha G (2021) A novel self-tuning fuzzy logic-based PID controllers for two-axis gimbal stabilization in a missile seeker. Int J Aerospace Eng 2021(1):8897556
19. Senthil Kumar S, Anitha G (2023) Fuzzy logic-based self-tuning PID controllers using parameters adaptive method for stabilization of a two-axis seeker gimbal. IETE J Res 69(9):6188–6197
20. Emamifard A, Ghadiri H (2023) Robust control of nonlinear fractional-order systems with unknown upper bound of uncertainties and external disturbance. IETE J Res 69(5):2621–2632
21. Ghadiri H, Khodadadi H, Mobayen S, Asad JH, Rojsiraphisal T, Chang A (2021) Observer-based robust control method for switched neutral systems in the presence of interval time-varying delays. Mathematics 9(19):2473
22. Shi Q, Wang H, Cheng H, Wang M, Cheng F, Liu F (2021) Fuzzy disturbance observer-based modified variable gains super-twisting sliding mode composite control for a two-axis airborne optoelectronic stabilized platform. Optik 228:166132
23. Ye R, Yan B, Shi K, Chen M (2020) Interval type-2 fuzzy sliding-mode control of three-axis stabilization gimbal. IEEE Access 8:180510–180519

24. Jamshidi S, Safa A, Mirzajani S (2025) Continuous nonsingular fast terminal sliding mode control for a free gyro stabilized mirror system. IEEE Trans Autom Sci Eng 22:10937
25. Yan M, Mou X, Wang R, Hu J (2025) Improved sliding mode control for airborne electro-optical stabilized platforms based on disturbance observer. In: 2025 8th international conference on advanced algorithms and control engineering (ICAACE). IEEE, pp 491–495
26. Khodadadi H, Ghadiri H (2019) Fuzzy logic self-tuning PID controller design for ball mill grinding circuits using an improved disturbance observer. Min Metallur Explor 36(6):1075–1090
27. Khodadadi H, Motlagh MRJ, Gorji M (2011) Robust control and modeling a 2-DOF inertial stabilized platform. In: International conference on electrical, control and computer engineering 2011 (InECCE). IEEE, pp 223–228
28. Heidarpoor S, Tabatabaei M, Khodadadi H (2017) Speed control of a DC motor using a fractional order sliding mode controller. In: 2017 IEEE international conference on environment and electrical engineering and 2017 IEEE Industrial and Commercial Power Systems Europe (EEEIC/I&CPS Europe). IEEE, pp 1–4
29. Ghadiri H, Khodadadi H, Eijei H, Ahmadi M (2021) Pso based takagi-sugeno fuzzy PID controller design for speed control of permanent magnet synchronous motor. Facta Universitatis Ser Electron Energetics 34(2):203–217
30. Samani R, Khodadadi H (2017) A particle swarm optimization approach for sliding mode control of electromechanical valve actuator in camless internal combustion engines. In: 2017 IEEE international conference on environment and electrical engineering and 2017 IEEE Industrial and Commercial Power Systems Europe (EEEIC/I&CPS Europe). IEEE, pp 1–4

Improving Hospital Building Thermal Comfort Prediction: A Machine Learning Approach

Hasibul Hasan Shawon, Mohammad Nyme Uddin, Ehetisum Sharif, Sabir Ahmed, Shaching Mong Marma, Muhammad Aziz Jobayer Hasan, and Muhammad Rasel Sheikh

Abstract Thermal comfort has a significant impact on productivity and occupant well-being. It has a meaningful effect on patient recovery, satisfaction, and overall health in hospitals. Naturally ventilated hospital buildings (NVHBs) are particularly sensitive environments due to the absence of mechanical cooling systems, even though they offer vital Medicare services to people from all socioeconomic backgrounds. This study developed machine learning (ML) techniques to accurately pre-

---

H. H. Shawon
Department of Civil Engineering, College of Engineering and Technology, IUBAT—International University of Business Agriculture and Technology, Dhaka, Bangladesh

Building Energy & Environmental Management (BEEM) Enhanced by AI, Dhaka, Bangladesh
e-mail: 22106159@iubat.edu

M. N. Uddin (✉)
Building Energy & Environmental Management (BEEM) Enhanced by AI, Dhaka, Bangladesh

Department of Architectural Engineering, College of Engineering, Imam Mohammad Ibn Saud Islamic University (IMSIU), Riyadh, Saudi Arabia
e-mail: mnuddin@imamu.edu.sa

E. Sharif
Building Energy & Environmental Management (BEEM) Enhanced by AI, Dhaka, Bangladesh

Department of Computer Science and Engineering, College of Engineering and Technology, IUBAT—International University of Business Agriculture and Technology, Dhaka, Bangladesh

S. Ahmed · S. M. Marma
Department of Civil Engineering, College of Engineering and Technology, IUBAT—International University of Business Agriculture and Technology, Dhaka, Bangladesh

Building Energy & Environmental Management (BEEM) Enhanced by AI, Dhaka, Bangladesh

M. A. J. Hasan
Department of Electrical and Computer Engineering, NSU—North South University, Dhaka, Bangladesh
e-mail: muhammad.hasan05@northsouth.edu

M. R. Sheikh
Department of Civil Engineering, World University of Bangladesh, Dhaka, Bangladesh

M. Mirrashid et al. (eds.), *Optimization and Adaptive Control Algorithms*, Emerging Trends in Mechatronics, https://doi.org/10.1007/978-981-95-8387-4_9

dict thermal comfort in NVHBs. Field data were collected from 200 inpatients in a rural NVHB in Faridpur, Bangladesh, using surveys, environmental monitoring, and patient feedback. The key features were selected by recursive feature elimination with cross-validation (RFECV) to improve model performance, and class imbalance was corrected using a hybrid of the edited nearest neighbor (ENN) approach and the synthetic minority over-sampling technique (SMOTE) method. This study presents the performance of three ML models: random forest (RF), decision tree (DT), and extreme gradient boosting (XGBoost), trained and evaluated using accuracy, precision, recall, and F1-score. Here, accuracy is a widely used and intuitive metric in classification models. RF outperforms all other models with a 98% accuracy on this dataset. The accuracies of XGBoost and DT were 95% and 91%, respectively. RFECV identified optimal features, and the combination of SMOTE-ENN balanced misclassification rates across different thermal comfort predictors. These results highlight both the limitation imposed by imbalanced datasets and the strengths of ML-based, interpretable models that can reliably predict thermal comfort in NVHBs, offering practical guidance for climate-resilient hospital design and improved patient well-being in resource-limited settings.

Keywords Thermal comfort · Naturally ventilated hospital building (NVHB) SMOTE-ENN · Imbalance classification · Rural area

1 Introduction

Thermal comfort in hospital buildings plays a pivotal role in patient recovery, satisfaction, and overall well-being, especially in rural regions where the healthcare infrastructure is often limited. In rural areas of developing countries like Bangladesh, healthcare infrastructure often relies on naturally ventilated buildings (NVHBs) due to financial constraints and a lack of mechanical cooling systems. Naturally ventilated buildings are typical in hot, humid regions like Madhukhali, Faridpur. But these types of buildings often fail to maintain optimal thermal conditions, potentially hindering patient comfort and recovery. Fanger's predicted mean vote (PMV) and other traditional thermal comfort models primarily rely on constant environmental elements, such as temperature and humidity. However, they overlook the diverse needs of patients, their medical conditions, and adaptive behaviors. This challenge highlights the pressing need for data-driven, context-sensitive solutions that better capture the intricacies of thermal perceptions in rural healthcare settings with natural ventilation [1, 2].

This study introduces three machine learning (ML) classification-based approaches using interpretable models, random forest (RF), decision tree (DT), and XGBoost, to predict patient thermal comfort in NVHB in rural environments. A variety of environmental, demographic, and architectural characteristics are incorporated into the study, including CO_2 levels, humidity, types of disease, length of hospitalization, distance from fans or windows, and surface temperature.

Additionally, patient-specific adaptive behaviors, such as clothing level and seating position, are considered. For feature selection, recursive feature elimination with cross-validation (RFECV) is used to identify the principal indicators. These variables are then used to train the models and evaluate classification performance using metrics such as accuracy, precision, recall, and F1-score. Similar ML approaches in thermal comfort studies have demonstrated the value of integrating physiological, behavioral, and sensor-based inputs to increase the accuracy of thermal comfort predictions [3].

These observations are the main objectives of this study, which include identifying the significant environmental, behavioral, and demographic factors affecting comfort levels as well as developing and validating interpretable ML models for classifying patient thermal comfort in a rural hospital. In the end, this study offers a strong foundation for optimizing patient care, recovery outcomes, and energy use in hot-humid climates through sustainable design and thermal management of rural hospital buildings. Ultimately, this study aims to identify key comfort factors and develop interpretable ML models to enhance patient care and energy efficiency in naturally ventilated rural hospitals in hot-humid climates.

2 Literature Review

Inclusive research has explored how occupants interact with indoor conditions and the role of natural ventilation in real-world situations. It shows that indoor conditions can vary greatly depending on occupants' actions such as opening windows or adjusting shades [4]. Theories on adaptive thermal comfort highlight that comfort levels often depend on cultural habits, personal routines, and local climate. Other studies indicate that poor indoor environments can negatively affect occupant well-being [5]. Thermal discomfort can reduce productivity and even lead to health problems [6]. These insights highlight the importance of comfort strategies that prioritize occupants and consider their specific contexts, especially in rural hospitals where resources are limited and natural ventilation is the primary method of cooling [7].

Thermal comfort is a crucial part of indoor environmental quality because it directly affects occupant well-being, productivity, and health, especially in sensitive places like hospitals [8]. For many years, standard models such as the PMV and the adaptive comfort model have provided a basis for estimating occupant comfort [9]. Another previous study has extensively explored thermal comfort using traditional simulation-based or statistical approaches. For instance, the American Society of Heating, Refrigerating and Air Conditioning Engineers (ASHRAE) global thermal comfort database II has been widely used to analyze PMV–Predicted Percentage of Dissatisfied (PPD) models across diverse climates and populations [10]. Similarly, thermal-radiative performance was evaluated using the ENVI-met model to assess the effectiveness of green infrastructure typologies in subtropical climates [11]. Another notable example is a simulation-based study that examined thermal comfort and energy demand in buildings in sub-Himalayan eastern India [12]. However, these models rely on fixed assumptions and general averages that may not perform well in naturally ventilated spaces such as rural hospitals. In contrast to controlled indoor settings, rural hospitals must adapt to varying weather and diverse occupant behaviors, which can reduce the accuracy of conventional models. Applying ML can help to overcome these limitations. ML shows its ability to capture complex indoor-human interactions, enabling more accurate and adaptive control of energy, air quality, and thermal comfort [13].

ML has become a significant tool for predicting thermal comfort because it can capture the complex ways occupants interact with their surroundings. Several popular ML methods such as RF, DT, and XGBoost have achieved outstanding results. For example, Chai et al. [14] accomplished over 85% accuracy using RF and gradient boosting to predict thermal comfort in naturally ventilated residential buildings. Katić et al. [2] improved personalized thermal comfort predictions by combining body measurements with occupants' heating use, reaching more than 80% accuracy. Liu et al. [15] combined different data sources and used ML with digital models of occupants to predict outdoor thermal comfort, showing very accurate results. However, challenges remain; these models require large, high-quality datasets, which are often difficult to obtain in rural hospital settings. Most studies focus on urban settings and residential buildings, with little attention given to hospitals, particularly rural ones, that rely on natural ventilation. Uddin et al. [16] recently pointed out this gap by applying ML models to predict thermal and visual comforts of occupants for a naturally ventilated educational building (NVEB) in low-income economies. Similar research is fundamental in the healthcare sector, where indoor conditions are greatly influenced by how patients use hospital spaces and options for maintaining thermal comfort are often limited.

This study aims to address an existing research gap by applying and combining RF, DT, and XGBoost algorithms to predict thermal comfort in NVHB, which is located in a rural region. RF and DT are appreciated for their clear interpretation and resistance to overfitting, while XGBoost provides higher accuracy through gradient boosting and handles complex nonlinear relationships effectively [14]. Adapting ML models in this setting can help to improve hospital design and daily operations, making the environment more comfortable for occupants [17]. The significance of this study lies in improving thermal comfort modeling for rural healthcare settings, which are currently underexplored in existing research. Better prediction accuracy helps to develop adaptive ventilation strategies and environmental controls, which are essential for NVHBs. These improvements help support occupant well-being and align with broader sustainability goals. By using advanced ML techniques, the study also contributes to global efforts to improve indoor environments and reduce energy consumption in healthcare facilities [18].

3 Methodology

The research methodologies were organized into three phases shown in Fig. 1. Phase I consists of an on-site survey that selects the case study in a hospital building and collects information from patients, including both quantitative and qualitative data collection processes. Additionally, the data curation and cleaning procedures

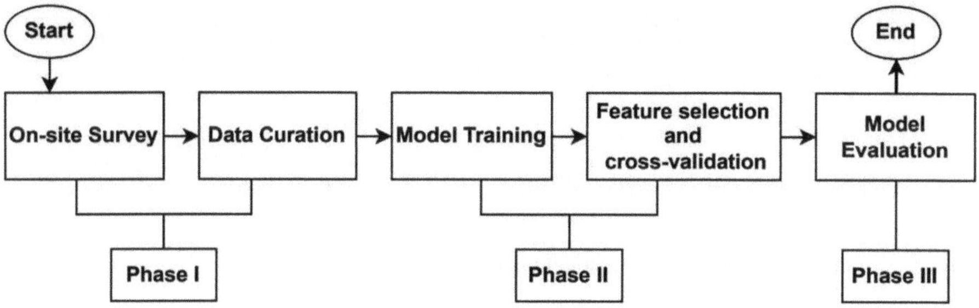

Fig. 1 Flowchart of research methodology

are required for subsequent analysis. Phase II covers the model training process, including parameter selection, data splitting, handling imbalanced classes, cross-validation, and final feature selection methods. Finally, phase III focuses on assessing the model's evaluation and checking prediction accuracy.

3.1 On-Site Survey

3.1.1 Case Study Building and Location

This study was conducted in a naturally ventilated hospital building (NVHB) located in the Madhukhali Upazila Health Complex. This healthcare facility is situated in Madhukhali, a rural administrative sub-district of the Faridpur district in Bangladesh, which experiences a hot and humid tropical climate. As the region's primary government hospital for the region, the complex plays a central role in meeting the healthcare needs of surrounding communities. The health complex was established in 1996 as a 31-bed facility on a site measuring approximately three acres. To expand its service capacity, a new four-story building was constructed in 2015 to increase the total bed capacity to 50. Although the upgraded structure was inaugurated in 2017, its intended operation has been hindered by severe shortages of both staff and financial resources. Currently, only the original 31 beds are in regular use, and nearly half of the 59 sanctioned staff positions remain vacant. Despite these constraints, the hospital remains a critical healthcare provider for the area, which offers both routine medical services and emergency care. Its location along the busy Dhaka–Khulna highway also positions it as a vital emergency response point for road traffic accidents. The facility has demonstrated resilience during public health crises, continuing to admit and treat large numbers of patients despite infrastructure and logistical challenges. These circumstances underscore the hospital's significance to the local healthcare system while also highlighting the urgent need for improved staffing levels, full utilization of infrastructure, and greater resource allocation to achieve its designed operational capacity. The specific NVHB examined in this study is a two-story hospital building with a total floor area of approximately 4600 square feet. It houses a range of functional spaces, including consultation rooms, a canteen, laboratories, doctors' and nurses' offices,

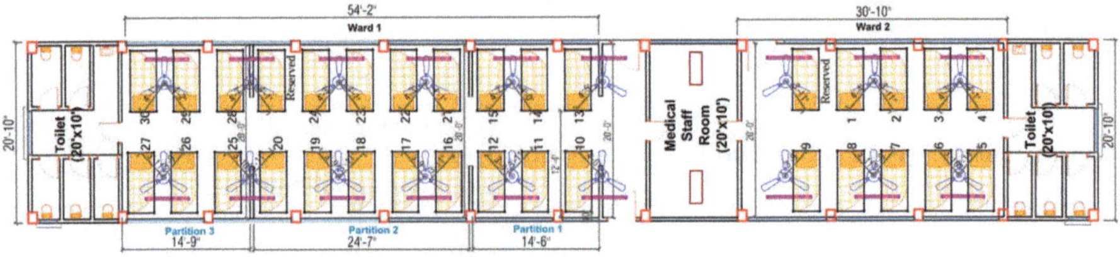

Fig. 2 Architectural layout of the hospital ward

an emergency department, and other essential support services. The ground floor is dedicated to outpatient care, whereas the upper floor accommodates inpatient wards. The second floor provides 30 general beds for admitted patients, along with two additional beds located in separate wards for specialized needs. For this research, only inpatients residing on the second floor were included in the survey. A schematic layout of the NVHB's inpatient wards is presented in Fig. 2.

3.1.2 Questionnaire Survey

The questionnaire survey comprised two main sections. The first section focused on patient demographics data, including age, gender, type of disease, and duration of hospital stay. The second section focused on various environmental and architectural aspects of the hospital ward, including floor area, number of fans and lights, distance between the patient's bed and the nearest window, as well as key indoor environmental parameters. These parameters included air temperature, surface temperature, relative humidity, carbon dioxide (CO_2) concentration, particulate matter ($PM_{1.0}$, $PM_{2.5}$, and PM_{10}), formaldehyde (HCHO), and total volatile organic compounds (TVOCs). This section also evaluates the patient's thermal comfort within the NVHBs. Participants were asked to rate their thermal sensation using the ASHRAE seven-point scale that indicates the range from cold (−3) to hot (+3), aligning with the values predicted by Fanger's predicted mean vote (PMV) model [19]. To supplement the quantitative data, the questionnaire incorporated open-ended questions aimed at capturing nuanced, qualitative insights from the patients. These questions provided participants with the opportunity to elaborate on their comfort experiences, identify specific sources of discomfort, suggest potential improvements to ward design or operations, and share personal narratives regarding their stay. The inclusion of such qualitative feedback was intended to provide a deeper, more context-rich understanding of patient comfort that may not be fully captured by numerical scales alone.

3.1.3 Data Collection

The survey was questionnaire based and actively conducted during the summer season from May 10 to July 1, 2025. The sample disciplines for the study were 200 male and female patients who were admitted to the inpatient wards located on the second floor of the hospital. The data were collected during the time frame of 11:30–17:00, as this period is when temperatures are highest during the summer season. The qualitative assessment was conducted through a Google Forms-based questionnaire, which collected qualitative environmental data using an innovative sensing instrument. The multiparameter was used to monitor indoor air quality while simultaneously measuring all or a combination of the following parameters: temperature, humidity, CO_2, $PM_{1.0}$, PM_{10}, $PM_{2.5}$, HCHO, and TVOC. On the other hand, light intensity was measured using a digital lux meter (model: AS803, Smart Sensor), and surface temperature was measured using an infrared thermometer (model: AR320, Smart Sensor). Four smart meters were installed across a critical building location to ensure spatial consistency and accuracy. Moreover, supplementary data on outdoor environmental data were used to correlate the external climatic conditions of NVHBs internally.

3.2 *Data Preparation*

Data curation refers to the process of selecting, organizing, and maintaining the items in a collection. This stage involves preparing the acquired data for analysis and cleaning it by removing missing values, outliers, and any discrepancies. Proper data processing is crucial for reliable and accurate thermal comfort prediction using ML models. A complete data cleaning process was used to clean the raw data and ensure that the dataset was clean. The cleanup phase involved identifying and addressing missing values, outliers, and inconsistent data entries. If missing values were found, they were sometimes replaced with the nearest values or removed manually. Outliers were detected using the box plot method and appropriately treated or removed to enhance model stability and generalization. Additionally, data structure inconsistencies such as misspellings, incorrect word usage, improper capitalization, and inconsistent naming conventions were corrected to improve data uniformity. Variance inflation factor (VIF) analysis was performed to identify and mitigate multicollinearity among predictor variables, ensuring that only relevant and independent features were used for model training. The min–max scaling strategy was used to standardize all numerical features to a comparable range suitable for ML techniques, thereby maximizing model performance. After dividing the dataset into 80% training and 20% test subsets, all preprocessing procedures, such as feature transformation and normalization, were applied to prevent data leakage. After this, only 200 records were left from the original dataset of 239. Data analysis was performed using a publicly accessible interactive cloud-based tool, Jupyter Notebook v7.2.2, to support recordkeeping, programming processes, and model visualization. It provided an environment that enabled a collaborative and transparent workflow to efficiently process the data from iterative steps in a reproducible way.

3.2.1 Approaches to Imbalanced Dataset Processing for Classification

The questionnaire survey datasets of thermal comfort are highly imbalanced over different categories, as illustrated in Fig. 3. There are only 2 respondents in the thermal comfort distribution in the “Cold” category, whereas the “Hot” category has

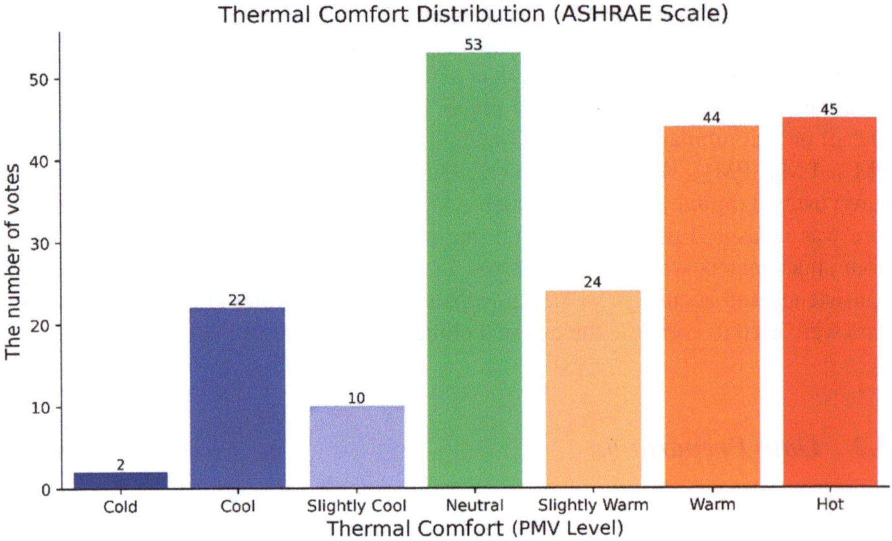

Fig. 3 Imbalance classification shown in thermal sensation

45 respondents. Similarly, the "Cool" class has only 22 respondents, while the "Warm" class has twice as many as the "Cool" class, that is, 44 respondents. Also, "Slightly Cool" has 10 respondents, while "Slightly Warm" has 24 respondents. The highest number of respondents is in the "Neutral" category, which is 53. These class distributions are typically imbalanced in multi-class classification, making this a challenging problem for developing models.

In classification tasks with imbalanced datasets, models tend to be biased toward majority classes and are often quick to misjudge the minority class. This is because minority class examples are rare and tend to have low predicted probabilities, which leads the model to be heavily influenced by false negatives rather than true positives during training. Some bootstrapping-based oversampling techniques solve this imbalance by generating more samples for the minority class. In simple words, minority examples are duplicated until they have a representation next to themselves enough times in the background data. This way expands the minority class dataset, and then in this manner, it helps in giving equal attention to all minority classes and primary data during the training of your model. One approach is the SMOTE, which addresses class imbalance by creating new, synthetic examples for the minority class rather than merely replicating the existing ones. This results in an improved generalization and classification accuracy for minority class instances, which mitigates the majority class bias. Table 1 presents several recent scholarly works that examine oversampling and resampling techniques, such as bootstrapping, for addressing imbalanced classification issues, including this study, which employs SMOTE-ENN for enhancing data balance and improving model robustness.

Table 1 Different types of oversampling and resampling techniques

Method	Key innovation	References
Kernel-based enhanced SMOTE	Kernel-weighted enhancement of SMOTE to generate synthetic samples	[20]
Comprehensive evaluation of SMOTE variants	Benchmarking SMOTE variants on transformer-embedded data	[21]
Hybrid clustering based oversampling and undersampling (HCBOU)	Hybrid cluster-based combination of oversampling and undersampling	[22]
Oversampling borderline minority instances (OBMI)	Targeted oversampling of borderline minority samples with Tomek Link cleaning	[23]
Synthetic minority oversampling technique combined with edited nearest neighbor (SMOTE-ENN)	For enhancing data balance and improving model robustness	This study

3.2.2 Synthetic Minority Oversampling Technique Combined with Edited Nearest Neighbor (SMOTE-ENN)

One of the techniques used for data cleaning is the ENN method, which applies the nearest neighbor rule to identify and remove any samples that are outliers concerning their surrounding data points. More precisely, it removes all observations that are misclassified according to its neighbors, which reduces noise and improves data quality. Combined with SMOTE, this approach benefits from SMOTE's ability to create synthetic samples for minority classes, along with ENN's data purifying capabilities.

3.3 Model Training

In this study, three machine learning models, DT, RF, and XGBoost, were employed to predict thermal comfort within the complex hospital environment characterized by numerous critical and interrelated parameters. These models were selected for their robustness, high interpretability, and strong ability to handle nonlinear relationships. The dataset was divided into 80% for training and 20% for testing to ensure effective model learning and reliable performance evaluation.

3.3.1 Random Forest (RF)

RF is an accurate ensemble learning algorithm in many varieties of data, going beyond assumptions of a simple parameterization. The ability to work with complex, nonlinear relationships and manipulate data alone must be considered one of several qualities that have given RF its reputation. It is suitable for imbalanced datasets as it can leverage class weighting or balanced sampling strategies to mitigate the majority-class bias. In addition, RF provides feature importance to help rank significant covariates; this is particularly appealing for thermal comfort modeling in which environmental parameters (e.g., air temperature, humidity, and air velocity) interact with personal factors (e.g., age and clothing level). It reduces overfitting by averaging predictions across multiple trees. Its nonlinear, intuitive, and strong generality characteristics fit multiple climatic and occupant situations in the most accessible way, widely employed to predict thermal comfort. RF prediction equations are presented in Eqs. (1) and (2) [24], for a dataset with N decision trees $T_1, T_2, \ldots, T_N$:

Classification:

$$\hat{y} = \text{mode}\left\{T_1(x), T_2(x), \ldots, T_N(x)\right\} \tag{1}$$

Regression:

$$\hat{y} = \frac{1}{N}\sum_{i=1}^{N} T_i(x) \tag{2}$$

where

- $T_i(x)$ = prediction from the ith decision tree for input x.
- $\hat{y}$ = final predicted class from the ensemble.
- N = total number of base learners in the ensemble.

3.3.2 Decision Tree (DT)

DT has the characteristics of transparency and simplicity, which are suitable for scenarios where good interpretability is paramount to approval, such as thermal comfort prediction applications. DT classifiers are a type of supervised learning algorithm that recursively splits a dataset based on feature values to predict categorical outcomes. The algorithm selects the feature and threshold at each node to maximize class purity for the split, using criteria such as Gini Impurity or Information Gain. This is recursively invoked for each of the nodes on which the split is made, until some stopping conditions like maximum tree depth or minimum samples per leaf are reached. For any input, DT routes the data through decision nodes to a leaf, which assigns the majority class label as the prediction. But DT has a strong tendency to overfit and is sensitive to small data changes, which could be handled via pruning and an ensemble method. DTs for classification operate by recursively splitting data into subsets based on feature values to predict classes, as shown in Eqs. (3), (4), and (5) [25].

The Gini Impurity at a node t is defined as

$$\text{Gini}(t) = 1 - \sum_{i=1}^{C} p_i^2 \tag{3}$$

where C is the number of classes and p is the proportion of samples of class i in node t.

Alternatively, the Information Gain from a split s is calculated as

$$IG(s) = \text{Entropy}(\text{parent}) - \sum_{j \in \{\text{left,right}\}} \frac{N_j}{N} \text{Entropy}(j) \tag{4}$$

where entropy at a node t is

$$\text{Entropy}(t) = -\sum_{i=1}^{C} p_i \log_2 p_i \tag{5}$$

and N and N_j denote the number of samples in the parent and child nodes, respectively. Splitting continues until stopping criteria such as maximum depth or minimum samples per leaf are met. For a new input, the tree assigns the class label of the majority in the leaf node corresponding to it. The transparent structure of DTs makes them especially suitable for domains that require interpretability, such as thermal comfort prediction. However, they may overfit training data, which can be mitigated via pruning or ensemble methods.

3.3.3 Extreme Gradient Boosting (XGBoost)

XGBoost is a powerful machine learning algorithm that improves speed and performance. It is a more advanced implementation of gradient boosting that builds an ensemble of decision trees sequentially, where each new tree tries to correct the errors made by the previous ones. What sets XGBoost apart is its emphasis on speed and performance through optimized parallel processing, handling of missing data, and regularization techniques to prevent overfitting. The model minimizes a custom objective function that combines a loss function measuring prediction error and a regularization term that controls model complexity. XGBoost has gained popularity for its scalability, accuracy, and flexibility, making it a favorite in competitions and practical applications like predicting thermal comfort, where it can learn complex relationships between environmental features and comfort levels. The multi-class classification model using XGBoost is formulated as follows in Eq. (6) [25]:

Probability for class k:

$$P(y_i = kx_i) = \frac{\exp\left(\sum_{t=1}^{K} f_t^{(k)}(x_i)\right)}{\sum_{j=1}^{C} \exp\left(\sum_{t=1}^{K} f_t^{(j)}(x_i)\right)} \tag{6}$$

where

- C = number of comfort categories.
- $f_t^{(k)}(x_i)$ = score from the t-th tree for class k.
- K = total number of base learners (iterations in boosting).

3.4 Feature Selection and Cross-Validation

Feature selection is a critical step in developing effective ML models, as it directly impacts model accuracy, complexity, and interpretability. The process aims to select and retain only the predictive factors significantly associated with the outcome and to drop unrelated or collinear features that could lead to noisy predictions or reduce the model's performance [26]. Dropping these features not only helps to alleviate overfitting but also reduces the related computational costs, making the model leaner and interpretable. One of the widely used and very effective feature selection approaches is recursive feature elimination (RFE). RFE is a wrapper-based technique that iteratively fits a model to the data, ranks the importance of features based on the model's coefficients or feature importance scores, and removes the least essential features at each iteration. The recursive approach continues until either several features are reached or the best subset is found. Now, the choice of a single data split made by RFE can lead to instability and overfitting. To alleviate this problem, RFE with cross-validation (RFECV) uses cross-validation as part of the RFE process. Consequently, it helps to increase the robustness and reliability of feature selection. In contrast to evaluating performance on a held-out validation set, RFECV performs RFE in a nested way and provides the optimal number of features to select from training-validation folds. This stratified cross-validation approach ensures that a particular data split does not bias the selection of features, thus improving the reproducibility and effectiveness of the chosen features [27].

In this study, RFECV was applied to three widely used ML algorithms, RF, DT, and XGBoost, to comprehensively evaluate the contribution of each feature toward predicting thermal comfort. These algorithms were selected due to their proven ability to handle complex and nonlinear relationships between predictors and the target variable, as well as for their differing mechanisms for assessing feature importance [24]. A stratified fivefold cross-validation in the RFECV was used to ensure unbiased evaluation during feature selection. Stratification ensures that an equal percentage of each label is presented in both the training and test sets, thereby preserving and balancing the proportions of thermal comfort classes. It helps prevent bias from imbalanced data distributions, leading to more accurate estimates of the model's performance. Initially, 30 features were considered as potential predictors based on previous domain knowledge and influence on thermal comfort parameters like air temperature, relative humidity, air velocity, and clothing level. The RFECV process was conducted to identify the narrowed-down list of most essential features found to be significant for thermal comfort prediction in different models and validation folds. This subset, which improved the predictive accuracy, also helped in creating models with higher interpretable values to better understand environmental and occupant-related variables that govern thermal sensation. Also, it can provide valuable knowledge for the development of adaptive building climate control strategies and personalized comfort systems that jointly try to achieve optimal indoor environmental quality and energy efficiency.

3.5 Model Evaluation

Evaluating ML models is a crucial step in the expedience and efficiency of these fast-evolving techniques to become reliable, robust, and valuable. Multiple metrics are needed, as simply stating one metric is crucial, which provides a multidimensional approach to understanding how well the model is performing. This is particularly true for classification-based ML models, where the goal is to categorize the input data accurately. The performance might be measured in precision, recall, F1-score, and accuracy depending on the faceted characteristics of a dataset in the predictive model. Accuracy is a widely used and intuitive metric in classification models. The overall accuracy is calculated as the total number of correctly classified instances out of the total samples. It is also defined as the ratio of correctly predicted observations to the total number of observations in the dataset. A higher accuracy value generally suggests better model performance, but accuracy alone can be misleading when dealing with imbalanced datasets where one class significantly outnumbers others [28, 29]. Under such conditions, the model may achieve high accuracy simply by predicting the majority class for all inputs, thereby failing to detect minority class instances effectively.

To address the limitations of accuracy, metrics such as precision and recall are utilized to provide deeper insights into the model's performance on specific classes, particularly the positive class in binary classification problems. Precision, also known as the positive predictive value, quantifies the proportion of accurate optimistic predictions among all instances that the model has predicted as positive. Also, precision tells us how many of the predicted positive cases were relevant [27]. High precision indicates that the model makes very few false-positive errors, which is crucial in scenarios where false alarms are costly or undesirable, such as spam detection or medical diagnosis. Recall is referred to as sensitivity or actual positive rate, which measures the model's ability to identify all relevant positive cases correctly. The proportion of accurate optimistic predictions over the total actual positive instances in the dataset is calculated. A high recall means that the model captures most of the positive cases and minimizes false negatives. This metric is critical in contexts where missing a positive case has serious consequences and fraud detection [30].

However, precision and recall often have an inverse relationship; improving one can lead to a decrease in the other. For example, a model that is tuned to have very high precision might become overly conservative, missing some true positives and lowering recall. On the other hand, a model with high recall might classify many instances as positive, with increasing false positives and reducing precision. To balance this trade-off between accuracy and recall, the F1-score is used. The F1-score is the harmonic mean of precision and recall, providing a single metric that captures the balance between these two essential aspects [31] because it weighs both false positives and false negatives equally. The F1-score is particularly useful when the class distribution is uneven or when one wants to find an optimal balance between identifying positive cases correctly and limiting false-positive errors.

The summary relies on a combination of metrics such as accuracy, precision, recall, and the F1-score, which offers a more nuanced and holistic assessment of a classification model's performance. Each metric highlights a different facet of effectiveness and enables practitioners to make informed decisions about model tuning. Also, it makes the selection and deployment based on the specific requirements and constraints of the problem domain [32].

The formulas for computing these metrics are presented in Eqs. (7)–(10) [33]:

$$\text{Precision} = \frac{TP}{TP + FP} \tag{7}$$

$$\text{Recall} = \frac{TP}{TP + FN} \tag{8}$$

$$F1 = 2 * \frac{\text{Precision} * \text{Recall}}{\text{Precision} + \text{Recall}} \tag{9}$$

$$\text{Accuracy} = \frac{TP + TN}{TP + FP + TN + FN} \tag{10}$$

Here, true positive (*TP*) refers to the number of correctly identified positive instances, false positive (*FP*) indicates the number of negative instances incorrectly classified as positive, true negative (*TN*) represents the correctly identified negative instances, and false negative (*FN*) denotes the number of positive instances that were incorrectly classified as negative.

4 Results and Discussion

Three ML algorithms, RF, DT, and XGBoost, were utilized to develop the classification models. To identify the most relevant input features, the RFECV technique was implemented. The model demonstrating the highest performance based on selected evaluation metrics was ultimately chosen as the most effective algorithm.

4.1 Feature Selection

The use of this method plays a crucial role in refining model inputs by identifying the most influential features and minimizing redundancy. This method enhances model generalization, lowers the risk of overfitting, and supports more reliable predictions across varied applications [34]. In this study, RFECV was applied separately to three classification models RF, DT, and XGBoost to identify the most significant features contributing to thermal comfort prediction. Figures 4, 5, and 6 present the feature selection process for each model, where the dotted line denotes the optimal number of features and the shaded band indicates the standard error during the resampling procedure. As shown in Table 2, each algorithm identified a different number of optimal features, highlighting how the model influences feature importance. XGBoost produced a compact set with nine features, whereas RF retained 24 and DT selected the most, totaling 27. These variations underscore the differing sensitivities of each model to input complexity and suggest that feature selection strategies should be tailored according to the specific algorithm used. Researchers have identified these factors as crucial for understanding and predicting thermal comfort. XGBoost's strong regularization and gradient-based feature evaluation make it more selective, retaining only the most influential predictors unlike RF and DT, which tend to include a broader set of variables. As an example, XGBoost

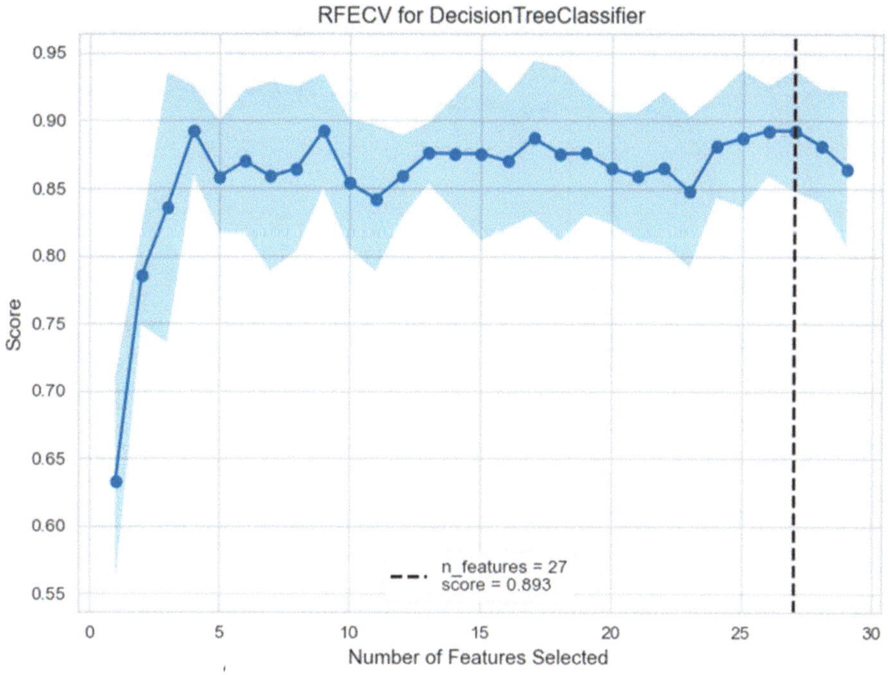

Fig. 4 RFECV with RF for thermal comfort prediction

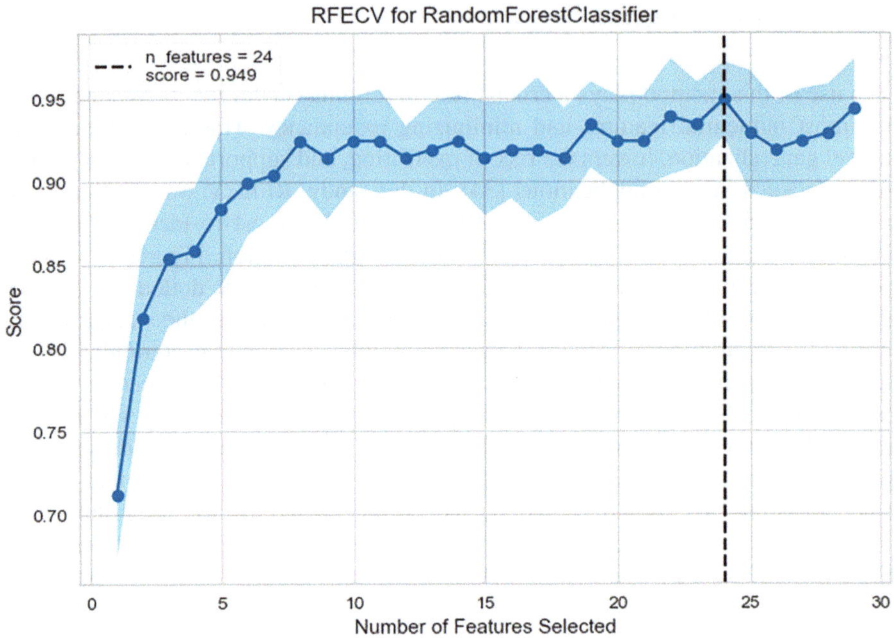

Fig. 5 RFECV with DT for thermal comfort prediction

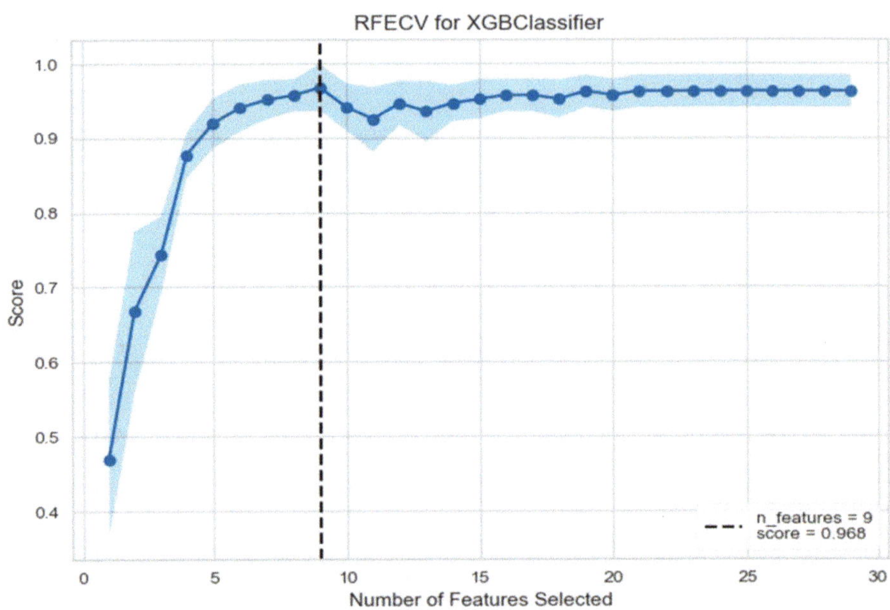

Fig. 6 RFECV with XGBoost for thermal comfort prediction

excluded features like bed number and window distance because they had little impact on prediction accuracy; window distance mattered less since windows were mostly closed, and bed number added minimal variability compared to key factors like temperature and air velocity.

The importance of several elements in thermal comfort modeling has been repeatedly emphasized in previous research, including gender, age [14, 35], window distance [36], ceiling fans [37], floor area [38], number of lights and lighting intensity [39], and room orientation [36]. These elements have been found to be crucial in determining and forecasting thermal comfort.

Table 2 Feature selection results for three different ML algorithms

Features	RF	DT	XGBoost
Gender			
Age	✓	✓	
Name_of_Diseases	✓	✓	✓
Admitted_Duration_(h)	✓	✓	
PMV_Level			
Visual_Comfort	✓	✓	
Clothing_Level	✓	✓	
Weather_Quality	✓	✓	✓
Bed_No	✓	✓	
Partition	✓	✓	
Nearest_Window_Distance_(ft)	✓	✓	
X-axis_Distance_From_Fan_(ft)	✓	✓	
No_of_ventilator		✓	✓
No_of_Fan		✓	
No_of_Window	✓	✓	
No_of_Light			
No_of_People_at_Each_Partition	✓	✓	✓
Partition_Area_(ft^2)	✓	✓	
Room_Area_(ft^2)		✓	
Surface_Temperature_(°C)	✓	✓	✓
Lighting_Level_(Lux)	✓	✓	
Temperature_(°C)	✓	✓	✓
Humidity_(%)	✓	✓	
Carbon-Di-Oxide_(CO_2)	✓	✓	
PM_2.5_($\mu g/m^3$)	✓	✓	
PM_1.0_($\mu g/m^3$)	✓	✓	
PM_10_($\mu g/m^3$)	✓	✓	✓
HCHO_(mg/m^3)	✓	✓	✓
TVOC_(mg/m^3)	✓	✓	
Wind_Velocity_(m/s)	✓	✓	✓

4.2 Model Performance

Table 3 presents the evaluation results of the three ML models, RF, DT, and XGBoost, which were developed using the combined sampling method of SMOTE and ENN. This particular strategy has proven more effective in handling imbalanced datasets than many conventional approaches [16]. All three models exhibited improved predictive performance compared to previously reported results. Among them, RF and XGBoost outperformed DT, achieving the highest accuracy and F1 scores, reinforcing previous findings that highlight the strong capabilities of these algorithms in sustainable built environment applications [40].

Beyond accuracy and F1-score, model evaluation also included precision and recall, analyzed separately for each model. In terms of precision, RF achieved the highest score, 0.98, closely followed by XGBoost, 0.96, while DT lagged slightly behind with a score of 0.93 under the ENN sampling condition. This finding is consistent with other studies [36] that also observed DT's comparatively poorer accuracy in categorizing thermal comfort levels. For recall, RF and XGBoost again showed comparable and robust performance, scoring 0.98 and 0.95, respectively. DT demonstrated a noticeably lower recall score of 0.91, further confirming its limitations in this context.

Additionally, a comprehensive performance evaluation of all classification models is summarized in Table 3, with accuracy and F1 score as primary indicators. These two metrics were selected because they provide a balanced view of model effectiveness by jointly considering precision and recall. Under these conditions, both RF and XGBoost achieved comparable and notably higher results in terms of F1 scores, 0.97 and 0.95, respectively, compared to the DT model's score of 0.91. A thermal comfort prediction model with an accuracy of about 80% was proposed in the study cited as [41]. This suggests that the model predicted thermal comfort with a reasonably high level of accuracy. Conversely, in this study, RF demonstrates a superior accuracy score of 0.98. XGBoost maintains its second position with an accuracy score of 0.95, while DT achieves the lowest accuracy score of 91. This indicates that the model was very successful in correctly predicting individual thermal comfort preferences. Accuracy and F1 score are good assessments for recall and accuracy. Because they consider precision and recall jointly, they provide a good sense of model performance.

The confusion matrices of the three machine learning models, RF, DT, and XGBoost, employed for thermal comfort prediction are presented in Figs. 7, 8, and 9, respectively. These metrics represent the model's performance in accordance

Table 3 Model performance of the three ML models

Model	Sampling strategy	Precision	Recall	F1-score	Accuracy
RF	SMOTE + ENN	0.98	0.98	0.97	0.98
DT	SMOTE + ENN	0.93	0.91	0.91	0.91
XGBoost	SMOTE + ENN	0.96	0.95	0.95	0.95

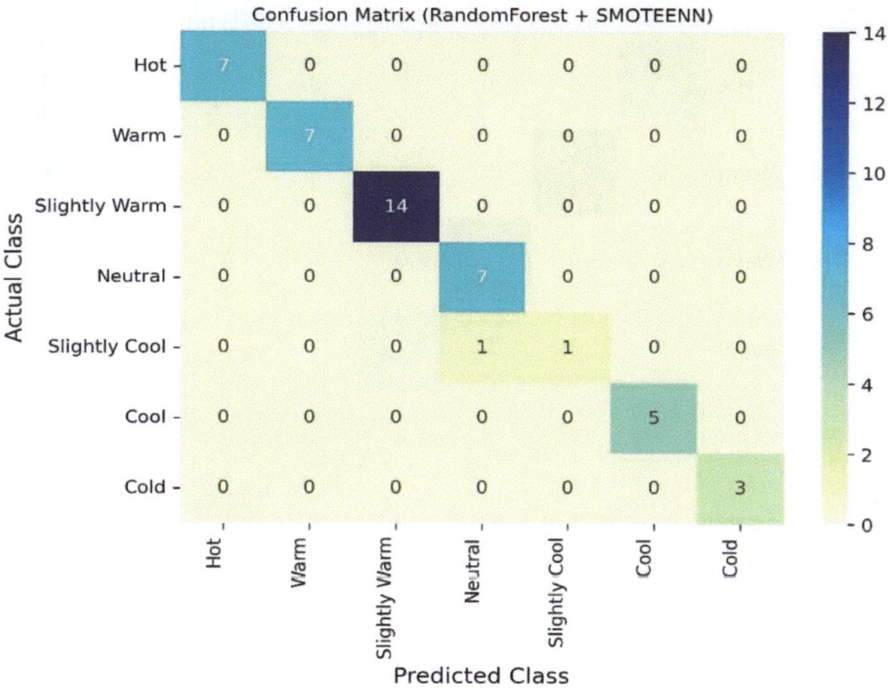

Fig. 7 Confusion matrix of RF for thermal comfort prediction

with the ASHRAE seven-point thermal comfort scale (ranging from +3 to −3), corresponding to the categories: Hot, Warm, Slightly Warm, Neutral, Slightly Cool, Cool, and Cold. The evaluation was conducted using the combined SMOTE and ENN sampling techniques. ENN improves model performance through several key mechanisms. It systematically eliminates noisy or mislabeled data points, reducing their potential to distort the learning process. Also, it removes instances that are positioned close to the minority class, which enhances the distinction between class boundaries and boosts prediction accuracy. The ENN data balancing method offers complete control over the data; it has been shown to be quite effective in achieving improved accuracy [42]. Its ability to integrate effectively with different classification algorithms further enhances model stability and contributes to a reduction in misclassification rates. To address the issue of class imbalance, the ENN technique was employed, which significantly improves the model's stability and overall performance.

The results of this study highlight the importance of improving predictive accuracy to gain a deeper insight into patients' responses, as well as the need for model interpretability to utilize it in practice. One of the main issues raised, which could affect model performance, is the problem of class imbalance. This was addressed by ensuring that the model training was balanced using appropriate data balancing techniques. Despite these approaches, some limitations still exist. The study utilized

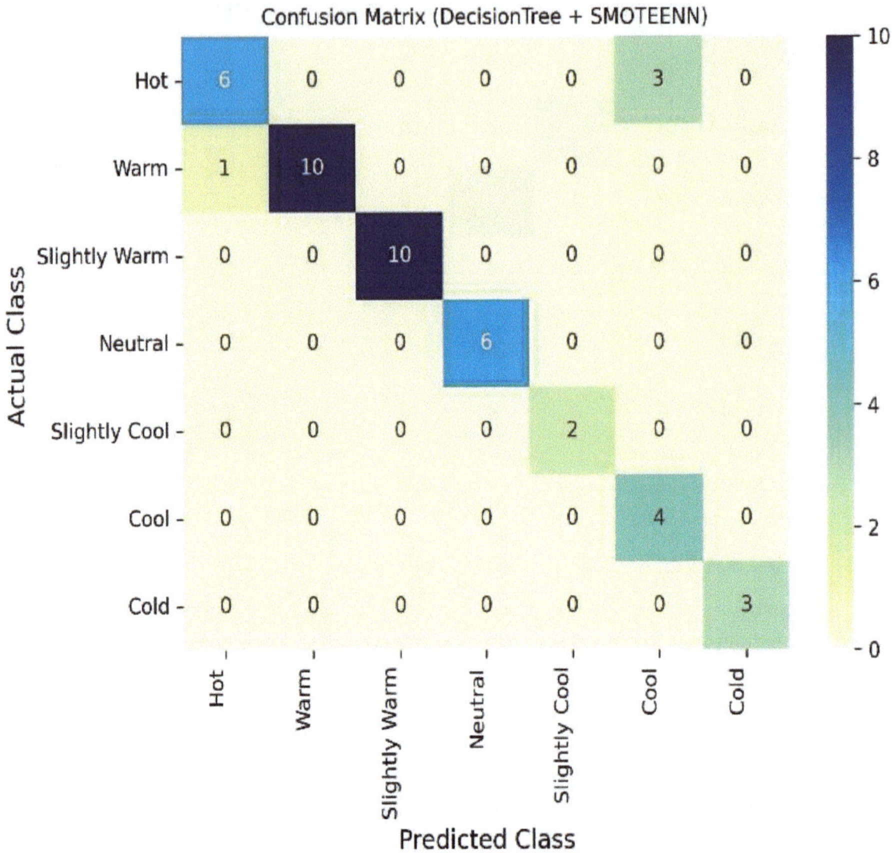

Fig. 8 Confusion matrix of DT for thermal comfort prediction

data obtained from a single NVHB in a rural area of Faridpur, Bangladesh, with a sample size of 200. These limitations may impact the generalizability of the findings to other types of buildings or geographical regions. Additionally, this study focuses on three ML algorithms, although other ML models may offer different viewpoints or better performance. For better applicability and more robust results discovered in future to be more generalized, more building types, larger datasets, and multiple study environments can be included in investigations.

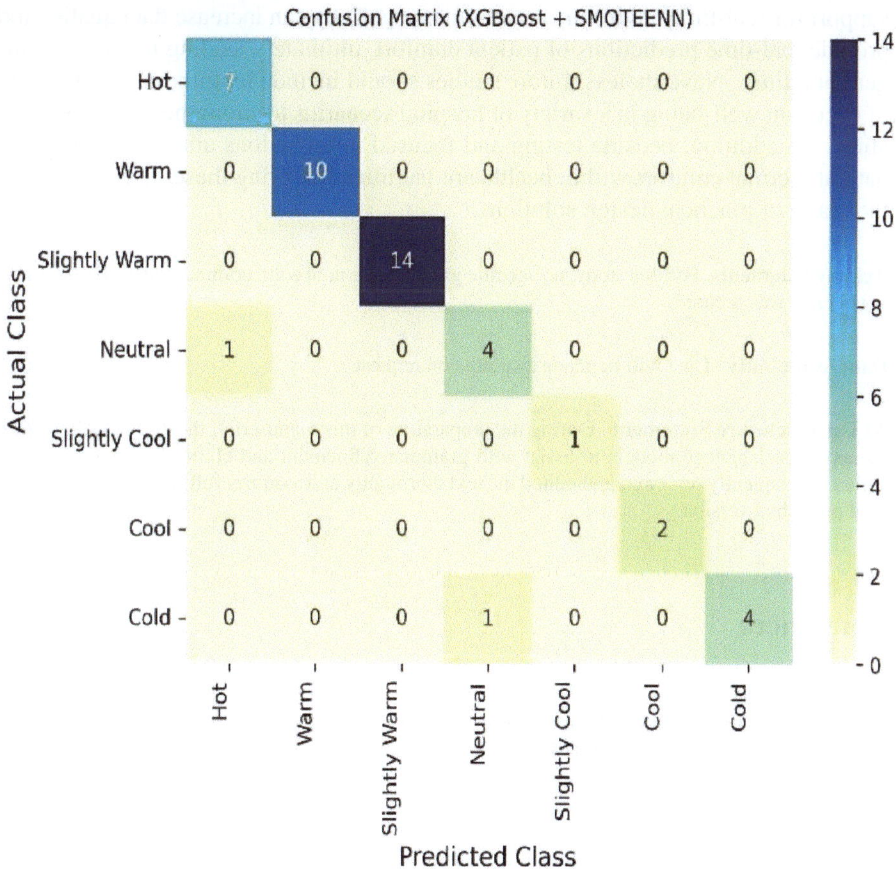

Fig. 9 Confusion matrix of XGBoost for thermal comfort prediction

5 Conclusion

This study emphasizes that machine learning can be effectively employed for thermal comfort prediction in NVHBs, and such predictive approaches may be especially valuable in rural regions like Faridpur, Bangladesh. The data were preprocessed to remove biases, and class imbalance was addressed using the SMOTE + ENN technique. Additionally, influential features were eliminated via RFECV, resulting in a robust framework for model training. Among the models, RF showed the highest prediction accuracy of 98%, followed by DT and XGBoost, which demonstrated good effectiveness in each case. These results indicate that NVHBs' thermal comfort scenarios can be accurately predicted by specific ML models, which are reportedly capable of effectively promoting occupant well-being inside their dwellings. Beyond the high predictive performance, this study sheds light on the value of innovative sensing technologies that IoT-based or portable environmental sensors can support for real-time environmental monitoring. They can increase data quality and provide real-time predictions of patient comfort, ultimately leading to more intelligent buildings. Nevertheless, future studies should include longitudinal monitoring of occupant well-being in a variety of hospital scenarios to strengthen the generalizability. In addition, bedside testing and focused interventions aimed at optimizing patient thermal comfort within healthcare facilities can bring these discoveries into the realm of practical design solutions.

Acknowledgments For this study, no specific grant from a nonprofit, commercial, or public funding agency was secured.

Data Availability Data will be made available on request.

AI Use Disclosure Statement During the preparation of this manuscript, the author utilized pre-trained large language models to assist with grammar refinement and clarity enhancement. The author subsequently reviewed and edited the text thoroughly and assumes full responsibility for the final published version.

References

1. Abd Rahman NM, Haw LC, Fazlizan A, Hussin A, Imran MS (2022) Thermal comfort assessment of naturally ventilated public hospital wards in the tropics. Build Environ 207:108480
2. Katić K, Li R, Zeiler W (2020) Machine learning algorithms applied to a prediction of personal overall thermal comfort using skin temperatures and occupants' heating behavior. Appl Ergon 85:103078
3. Ho T-F et al (2024) Thermal comfort model established by using machine learning strategies based on physiological parameters in hot and cold Environments. Indoor Air 2024(1):9427822
4. Roetzel A, Tsangrassoulis A, Dietrich U, Busching S (2010) A review of occupant control on natural ventilation. Renew Sust Energ Rev 14(3):1001–1013
5. Al Horr Y, Arif M, Katafygiotou M, Mazroei A, Kaushik A, Elsarrag E (2016) impact of indoor environmental quality on occupant well-being and comfort: a review of the literature. Int J Sustain Built Environ 5(1):1–11
6. Lan L, Wargocki P, Lian Z (2011) Quantitative measurement of productivity loss due to thermal discomfort. Energy Build 43(5):1057–1062
7. Singh MK, Mahapatra S, Atreya SK (2010) Thermal performance study and evaluation of comfort temperatures in vernacular buildings of North-East India. Build Environ 45(2):320–329
8. Frontczak M, Wargocki P (2011) Literature survey on how different factors influence human comfort in indoor environments. Build Environ 46(4):922–937
9. Nicol JF, Humphreys MA (2002) Adaptive thermal comfort and sustainable thermal standards for buildings. Energy Build 34(6):563–572
10. Cheung T, Schiavon S, Parkinson T, Li P, Brager G (2019) Analysis of the accuracy on PMV—PPD model using the ASHRAE Global Thermal Comfort Database II. Build Environ 153:205–217
11. Ouyang W, Sinsel T, Simon H, Morakinyo TE, Liu H, Ng E (2022) Evaluating the thermal-radiative performance of ENVI-met model for green infrastructure typologies: experience from a subtropical climate. Build Environ 207:108427

12. Thapa S et al (2023) Simulation of thermal comfort and energy demand in buildings of sub-Himalayan eastern India—impact of climate change at mid (2050) and distant (2080) future. J Build Eng 68:106068
13. Zhang W, Wu Y, Calautit JK (2022) A review on occupancy prediction through machine learning for enhancing energy efficiency, air quality and thermal comfort in the built environment. Renew Sust Energ Rev 167:112704
14. Chai Q, Wang H, Zhai Y, Yang L (2020) Using machine learning algorithms to predict occupants' thermal comfort in naturally ventilated residential buildings. Energy Build 217:109937
15. Liu X, Gou Z, Yuan C (2024) Application of human-centric digital twins: predicting outdoor thermal comfort distribution in Singapore using multi-source data and machine learning. Urban Clim 58:102210
16. Uddin MN et al (2024) Thermal and visual comforts of occupants for a naturally ventilated educational building in low-income economies: a machine learning approach. J Build Eng 94:110015
17. Sattayakorn S, Ichinose M, Sasaki R (2017) Clarifying thermal comfort of healthcare occupants in tropical region: a case of indoor environment in Thai hospitals. Energy Build 149:45–57
18. Hussien A, Maksoud A, Al-Dahhan A, Abdeen A, Baker T (2025) Machine learning model for predicting long-term energy consumption in buildings. Discov Internet Things 5(1):18
19. Uddin MN, Wang Q, Wei HH, Chi HL, Ni M (2021) Building information modeling (BIM), System dynamics (SD), and Agent-based modeling (ABM): towards an integrated approach. Ain Shams Eng J 12(4):4261–4274
20. Li W, Zhu S, Li Z, Wang H (2025) Kernel-based enhanced oversampling method for imbalanced classification. ArXiv
21. Taskiran SF, Turkoglu B, Kaya E, Asuroglu T (2025) A comprehensive evaluation of oversampling techniques for enhancing text classification performance. Sci Rep 15(1):21631
22. Salehi A, Khedmati M (2025) Hybrid clustering strategies for effective oversampling and undersampling in multiclass classification. Sci Rep 15(1):3460
23. Leng Q, Guo J, Tao J, Meng X, Wang C (2024) OBMI: oversampling borderline minority instances by a two-stage Tomek link-finding procedure for class imbalance problem. Complex Intell Syst 10(4):4775–4792
24. Breiman L (2001) Random forests. Mach Learn 45(1):5–32
25. Friedman JH (2001) Greedy function approximation: a gradient boosting machine. Ann Stat 29:1189–1232
26. Guyon I, Elisseeff A (2006) An introduction to feature extraction. In: Guyon I, Nikravesh M, Gunn S, Zadeh LA (eds) Feature extraction: foundations and applications. Springer Berlin Heidelberg, Berlin, pp 1–25
27. Powers DM (2020) Evaluation: from precision, recall and F-measure to ROC, informedness, markedness and correlation
28. Johnson JM, Khoshgoftaar TM (2019) Survey on deep learning with class imbalance. J Big Data 6(1):27
29. Saito T, Rehmsmeier M (2015) The precision-recall plot is more informative than the ROC plot when evaluating binary classifiers on imbalanced datasets. PLoS One 10(3):e0118432
30. Chicco D, Jurman G (2020) The advantages of the Matthews correlation coefficient (MCC) over F1 score and accuracy in binary classification evaluation. BMC Genomics 21(1):6
31. Sokolova M, Lapalme G (2009) A systematic analysis of performance measures for classification tasks. Inf Process Manag 45(4):427–437
32. Sun Y, Wong AKC, Kamel MS (2009) Classification of imbalanced data: a review. Int J Pattern Recognit Artif Intell 23(04):687–719
33. Yang B et al (2022) Comparison of models for predicting winter individual thermal comfort based on machine learning algorithms. Build Environ 215:108970
34. Yang B, Haghighat F, Fung BCM, Panchabikesan K (2021) Season-based occupancy prediction in residential buildings using machine learning models. e-Prime Adv Electr Eng Electron Energy 1:100003

35. Uddin MN, Lee M, Ni M (2023) The impact of socio-demographic factors on occupants' thermal comfort and sensation: an integrated approach using statistical analysis and agent-based modeling. Build Environ 246:110974
36. Zhuang D, Wang T, Gan VJL, Zhao X, Yang Y, Shi X (2022) Supervised learning-based assessment of office layout satisfaction in academic buildings. Build Environ 216:109032
37. Mihara K, Sekhar C, Takemasa Y, Lasternas B, Tham KW (2019) Thermal comfort and energy performance of a dedicated outdoor air system with ceiling fans in hot and humid climate. Energy Build 203:109448
38. Li J, Zheng B, Bedra KB, Li Z, Chen X (2022) Effects of residential building height, density, and floor area ratios on indoor thermal environment in Singapore. J Environ Manag 313:114976
39. Prasad SS et al (2025) Explainable hybrid deep learning framework for enhancing multi-step solar ultraviolet-B radiation predictions. Atmos Environ 343:120951
40. Martínez-Comesaña M, Eguía-Oller P, Martínez-Torres J, Febrero-Garrido L, Granada-Álvarez E (2022) Optimisation of thermal comfort and indoor air quality estimations applied to in-use buildings combining NSGA-III and XGBoost. Sustain Cities Soc 80:103723
41. Cosma AC, Simha R (2019) Machine learning method for real-time non-invasive prediction of individual thermal preference in transient conditions. Build Environ 148:372–383
42. Liu T, Jin L, Zhong C, Xue F (2022) Study of thermal sensation prediction model based on support vector classification (SVC) algorithm with data preprocessing. J Build Eng 48:103919

Metaheuristic Algorithms in Geotechnical Design: A Short Review on Optimising Shallow Foundations

Sahar Ebrahimi, Chia Yu Huat, Mohammadreza Koopialipoor, and Pejman Sabet

Abstract To ensure the safety and serviceability of the footing design, two fundamental criteria must be met: bearing capacity, representing the geotechnical ultimate limit state (ULS), and settlement, representing the serviceability limit state (SLS). There are specific requirements regarding these two criteria, such as achieving a factor of safety of 3 and ensuring that settlement does not exceed 25 mm for a simple pad footing design. On complying with this safety requirement, the cost of the footing plays an important role in design requirements, with material, dimensions, excavation and number of reinforcement bar components that affect the cost. Recent studies are increasingly using optimisation algorithms (OAs) to identify the optimum dimensions and materials for the construction of footings. The primary objective of this review is to optimise shallow foundation design, specifically focusing on the construction cost of the footing using OA techniques. In addition, the study provides recommendations on how to start with a simple pad footing design and use the OA in practice.

Keywords Shallow foundation · Optimisation algorithm · Design parameter Cost estimation

---

S. Ebrahimi
School of Civil Engineering, College of Engineering, University of Tehran, Tehran, Iran
e-mail: ebrahimi.sahar@ut.ac.ir

C. Y. Huat
Department of Civil Engineering, University of Malaya, Kuala Lumpur, Malaysia

M. Koopialipoor (✉)
Faculty of Civil and Environmental Engineering, Amirkabir University of Technology, Tehran, Iran
e-mail: Mr.koopialipoor@aut.ac.ir

P. Sabet
Civil Construction Department, Acknowledge Education, Melbourne, VIC, Australia
e-mail: Pejman.Sabet@ae.edu.au

M. Mirrashid et al. (eds.), *Optimization and Adaptive Control Algorithms*, Emerging Trends in Mechatronics, https://doi.org/10.1007/978-981-95-8387-4_10

1 Introduction

A shallow foundation is a component of the structural system designed to transfer loads directly to the soil during the process of soil-structure interaction. To ensure effectiveness, a shallow foundation must fulfil two key requirements: it must possess sufficient strength to resist applied loads without failure (ultimate limit state, ULS), and it must prevent excessive deformation or damage under the imposed loads (serviceability limit state, SLS) [1–3]. A shallow foundation is commonly adopted for low-rise structures, such as low-rise houses and bungalows [4, 5], provided that the ground strength is sufficient to withstand the loads.

In current geoengineering practice regarding design of shallow foundations, designs are mainly focused on satisfying ULS and SLS requirements. In terms of economic design, the conventional approach often relies on a trial-and-error process to determine different footing dimensions under similar ground conditions. This approach, however, is time-consuming and tedious. In recent years, many researchers have explored the use of optimisation algorithms (OAs) to enhance the design process by improving both accuracy and calculation efficiency in geotechnical engineering projects, including foundation design [6–9], tunnelling [10–12], pile bearing capacity [13–15], slope stability [16–21] and retaining wall [22–26].

There are many types of OAs available; however, the most common methods adopted in geoengineering studies are particle swarm optimisation (PSO) [27–32], artificial bee colony algorithm [33–36], imperialist competitive algorithm (ICA) [14, 37–39] and genetic algorithm (GA) [40–43].

This review chapter provides essential guidelines for researchers and engineers on integrating and applying OA techniques in footing design and analysis. First, the study reviews the main formulas for calculating the bearing capacity of the footing and discusses the relevant design parameters. Next, a review of recent research using OAs to determine and identify the optimal dimensions of the footing and their impact on calculations is presented. Finally, a recommended approach is proposed based on insights from previous research conducted on shallow foundations.

2 Shallow Foundation Design

Terzaghi [44] delineated a shallow foundation as a foundation for which the width, B, is equal to or less than the depth, D_f, and proposed the ultimate bearing capacity (UBC) of the shallow foundation via the subsequent formula:

$$q_{u=1.3cN_c+qN_q+0.4\gamma BN_\gamma} \tag{1}$$

for a square foundation of size $B \times B$

and

$$q_{u=1.3cN_c+qN_q+0.3\gamma DN_\gamma} \tag{2}$$

for a circular foundation with diameter *D*

where

c is the cohesion of the soil.
q is the surcharge, which can be defined as γd_f.
γ is the unit weight of the soil.
B is the width of the foundation.

Meyerhof [45] revised the formula to the following widely used formulation:

$$q_u = cN_c + qN_q + 0.5\gamma BN_\gamma \tag{3}$$

where

N_c, N_q and N_γ are dimensionless bearing capacity factors.

Other authors have also proposed formulas, aiming to ease the computation of the bearing capacity of the footing.

After computing the UBC of the shallow foundation, the factor of safety (FOS) can be computed as follows:

$$FOS = \frac{q_{ult}}{q_{\max}} \tag{4}$$

where q_{max} is the maximum allowable bearing pressure.

Most standards require the designed shallow foundation to have FOS equal to a minimum value of 3 [46–49]. Other than footing stability, footing settlement must also be checked as part of the SLS checking. There are several methods that can be used to determine the settlement of the footing, such as the finite element model (FEM) [50–53], which is highly accurate but time-consuming. By assuming that the soil behaves elastically, the footing foundation settlement can be calculated as follows as proposed by [54]:

$$\delta = \frac{P\left(1-v^2\right)}{\beta_z E\sqrt{BL}} \tag{5}$$

where

P is the applied load on the foundation.
v is the Poisson's ratio of the soil.
E is the elastic modulus of the soil.
β_z is the influence factor, which depends on the shape and flexibility of the foundation and the depth of embedment.

After checking ULS and SLS, various inputs for different dimensions can be added into the formula to work out various outputs, provided that these inputs comply with design codes, geotechnical parameters and material properties. The settlement equation typically incorporates factors such as the applied load, footing dimensions, soil stiffness (modulus of elasticity), Poisson's ratio of the soil and depth of the foundation. It is essential to ensure that all inputs are accurate and within the permissible limits specified by relevant standards to achieve reliable results. Additionally, the soil type and its behaviour under load, such as consolidation or immediate settlement, must be considered to ensure that the foundation performs as intended over its design life. Proper validation of inputs and outputs is crucial to avoid overestimation or underestimation of settlement, which could lead to structural instability or excessive deformation.

The reinforcement bar of the footing can be calculated based on the various thicknesses of the proposed footing. As such, the minimum reinforcement of the footing can be calculated as follows based on Eurocode 2 [55]:

$$\max\left(A_{s,\min} = 0.26\frac{f_{ctm}}{f_{yk}} \times B \times d, 0.0013bd\right) \tag{6}$$

where

f_{ctm} is the axial tensile strength of the concrete.
f_{yk} is the yield strength of the reinforcement.
B is the dimension of the footing.
d is the effective depth of footing.

For the shear reinforcement, a footing thickness of 250 mm or more is required to prevent shear failure. Footings carrying light column loads such as 100–400 kN on good soil may require only the minimum reinforcement. For footings supporting low-rise buildings of 1–2 stories with strong soil conditions of UBC more than 200 kPa, minimum reinforcement is often sufficient. This is demonstrated in the calculation below for the example information, as follows:

Footing size: 2.0 m × 2.0 m
Column load: 400 kN
Soil bearing capacity: 200 kPa
Concrete grade: 25 MPa
Yield strength of reinforcement: 500 MPa
Effective depth of footing: 0.35 m

Bending moment on the footing:

$$M = \frac{q \cdot L^2}{8} \tag{7}$$

$$M = \frac{200 \times 10^3 \times 2^2}{8} = 100\,\text{kN m}$$

Minimum reinforcement of the required steel reinforcement can be determined as follows:

$$A_{s,req} = \frac{M}{0.87 f_{yk} d} \tag{8}$$

$$A_{s,req} = \frac{100 \times 10^6}{0.87 \times 500 \times 0.35} = 656.8\,\text{mm}^2$$

Minimum reinforcement requirement:

$$A_{s,\min} = \rho_{\min} \times B \times d \tag{9}$$

$$A_{s,\min} = 0.0013 \times 2000 \times 350$$

$$A_{s,\min} = 910\,mm^2$$

Since the requirement of the steel reinforcement $A_{s,\,req}$ is less than $A_{s,\,min}$, the minimum reinforcement always governs the design for this low-rise building.

The above explanation is the conventional small footing size, which can be calculated. However, in actual construction, footing design can be controlled by various dimensions, different ground conditions, different loading and many other aspects.

After setting up the criteria for footing for the ULS and SLS checking, various inputs can be set so that OAs can be used to identify the optimum dimensions, material and size of the reinforcement to determine the optimum design. Table 1 shows the potential various dimensions that could be considered in the analysis.

According to [6], total material construction cost may be calculated as follows:

$$f_{cost} = C_e V_e + C_f A_f + \xi C_r M_r + \frac{f_c'}{f_{c\min}'} C_c V_c + C_b V_b \tag{10}$$

Table 1 Various dimensions and materials that can be used for footing design (adapted from [56])

Variable	Recommended range of the inputs
B (m)	0.5 to 5.0
L (m)	0.5 to 5.0
t (m)	0.4 to 2.0
Φ (mm)	8, 10, 12, 16, 18, 20
f_{ck} (MPa)	25, 30, 35, 40,45,50

Note: Φ is the diameter of the reinforced steel bar, f_{ck} is the concrete compressive strength

where

C_e is the unit cost of excavation per m^3.
C_f is the unit cost of the framework per m^2.
C_r is the unit cost of reinforcement per kg.
C_c is the unit cost of concrete per m^3.
C_b is the unit cost of backfill per m^3.
$f_c^{'}$ is the compressive strength of concrete.
$f_{c\min}^{'}$ is the minimum compressive strength of concrete.
V_e is the volume of excavation.
A_f is the area of the framework.
M_r is Reinforcement weight (kg).
V_c is the volume of concrete.
V_b is the volume of backfill.
ξ is the reduction factor.

3 Applications of OAs to a Wide Range of Shallow Foundation Problems

The following section conducts a thorough review of the various applications of OAs in shallow foundation design, highlighting how these soft computing techniques effectively address challenges. OAs are methods that can be used to solve problems that require minimising or maximising different conditions. Each problem has an objective function that is affected by a variety of parameters. OAs have been developed and applied by various researchers for the design of shallow foundations, taking into account different objectives, such as maximising the FOS, enhancing the bearing capacity, minimising construction costs and improving performance. There are a host of varieties of OAs, and depending on the problem characteristics and design conditions, a number of them can be implemented for a specific problem, and the results compared. Figure 1 illustrates a general flowchart that can be used in the design of shallow foundations that clearly shows which path was used to optimise the parameters.

In general, two strategies have been used to tackle footing problems. The first involves conducting a direct optimisation procedure, while the second employs a hybrid approach that integrates predictive and enabling models (e.g. artificial neural network, ensemble techniques, random forest, support vector machine) with optimisation techniques to optimise hyperparameters. The latest studies regarding the first strategy are discussed below.

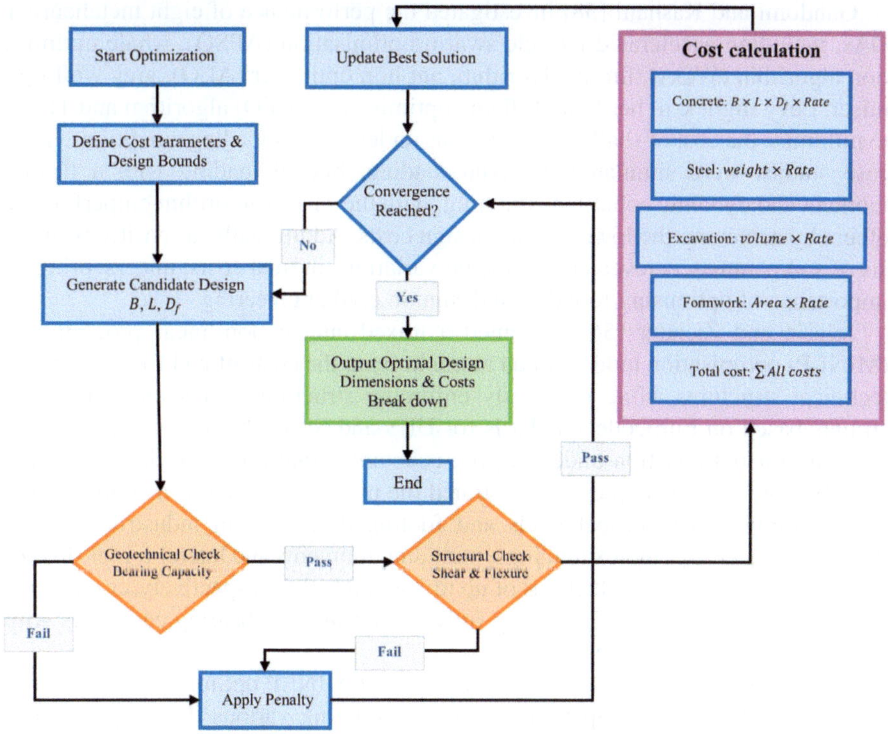

Fig. 1 Optimisation flowchart for minimum-cost pad footing design

3.1 Direct Application of OAs

Recently, much research has been conducted on the design of shallow foundations using a variety of OAs. Considering the numerous applications of OAs, an attempt has been made in this section to examine various aspects of these studies and highlight key points that may be useful for future research in this field.

Nigdeli et al. [57] proposed a metaheuristic-based methodology for evaluating the cost function of reinforced concrete (RC) footings, focusing on the optimisation of footing dimensions, column orientation and reinforcement design. The methodology employed several algorithms, including teaching-learning-based optimisation (TLBO), harmony search (HS) and flower pollination algorithm (FPA), to effectively address non-linear optimisation problems. The research emphasised the significance of design variables, including footing dimensions and loading conditions, in achieving optimal solutions that take into account both geotechnical and structural limit states, ultimately demonstrating the effectiveness of the proposed methods in reducing costs.

Gandomi and Kashani [58] investigated the performance of eight metaheuristic OAs, including accelerated particle swarm optimisation (APSO), whale optimisation algorithm (WOA), firefly algorithm, ant lion optimiser (ALO), grey wolf optimiser, Lévy flight krill herd, moth-flame optimisation (MFO) algorithm and TLBO, to minimise the cost of shallow foundations under various loading conditions. Three case studies were simulated: uniaxial loading, biaxial loading with a flexural moment and dynamic column positioning, with the TLBO algorithm outperforming others by achieving the lowest mean design costs. Additionally, a sensitivity analysis of soil parameters revealed significant variations in final costs, underscoring the importance of optimising foundation design in civil engineering.

Jelušič and Žlender [56] presented a mixed-integer non-linear programming (MINLP) optimisation model aimed at minimising the costs of pad footings in geotechnical structures while rigorously enforcing structural and geotechnical constraints based on Eurocode standards for ULS and SLS. The model was versatile, allowing for optimisation under varying economic conditions and design parameters. Two practical examples demonstrated the model's efficiency: a simple rectangular footing under vertical loads and footing design in an industrial concrete building subject to complex loadings, including moments and lateral forces. Results showed significant cost reductions of up to 36% and 63%, respectively, compared to initial design proposals, achieved by fully exploiting load-bearing capacities without unnecessary safety reserves.

Jelušič and Žlender [59] discussed a series of MINLP optimisations conducted on reinforced pad and strip foundations, considering various design parameters, including vertical loads, soil shear angles, cross sections and reinforcement layers. The study highlighted that structural optimisation is significantly more efficient than classical analysis, as it maximises the use of material without excess resistance. The model minimised the self-manufacturing cost per metre, incorporating material and labour expenses, while adhering to guideline and standard constraints for ULS and SLS. The findings recommended an optimal design for engineers to select suitable foundations based on specific vertical loads and soil shear strength, indicating that reinforced foundations are most cost-effective for high vertical loads on soils with low shear strength.

Islam and Rokonuzzaman [48] presented an optimised design process for shallow isolated column footings in sand using a GA to minimise total construction cost. The optimisation treated the footing width, length, thickness and depth as design variables, intending to minimise cost while satisfying geotechnical constraints: a minimum FOS against bearing capacity failure ($FS \geq 3.0$) and a maximum allowable settlement ($\delta \leq 25$ mm). The methodology included a tournament selection scheme for reproduction and a crossover process to enhance the optimisation process, addressing issues such as stagnation and premature convergence in design solutions, which led to potential cost savings of up to 68% compared to conventional designs.

Azhim and Prakoso [60] presented an optimisation approach for designing shallow foundations that simultaneously considers ULS, SLS and economic factors using the Solver add-in function in MS Excel. The research aimed to create a design methodology that explicitly incorporates cost aspects to achieve economically optimised foundation designs while ensuring compliance with ULS and SLS requirements. Through sensitivity analysis, the study identified key parameters, such as soil properties and work unit prices, that significantly affect construction costs, ultimately providing a framework for minimising foundation expenses across various Indonesian provinces.

Kashani et al. [6] evaluated the effectiveness of evolutionary algorithms, including differential evolution (DE), biogeography-based optimisation (BBO), evolutionary strategy (ES) and their variants [i.e. improved DE based on an adaptive mutation scheme, linear population size reduction success-history-based adaptive DE, weighted differential evolution algorithm (WDE) and BBO with covariance matrix-based migration], in optimising the cost of RC shallow footings per ACI 318-05 standards under various loading and design scenarios. DE, ES, BBO and improved versions were tested on three cases: uniaxial loading, combined axial and flexural loading and variable column location on the footing. BBO and WDE algorithms demonstrated superior capability in optimising shallow footing designs under complex loading and design conditions, outperforming traditional methods and other evolutionary variants. Sensitivity analyses confirmed footing depth, concrete strength and load inclination as critical cost drivers.

Chaudhuri and Maity [61] detailed the development and application of two optimisation algorithms—binary-coded GA with a static penalty function and unified particle swarm optimisation (UPSO)—for cost-effective design of isolated rectangular RC footings. The study incorporated Indian standard guidelines, focusing on minimising total footing cost by optimising design variables such as footing depth, width and steel reinforcement area. UPSO exhibited faster convergence and broader search capabilities, while GA offered conservative design advantages. Developed programs are adaptable to various isolated footing types and potentially extendable to combined, mat or pile foundations.

Debnath [62] examined the seismic bearing capacity of shallow strip footings situated on slopes on two-layered soil using a pseudo-static limit equilibrium analysis. They proposed a linear failure mechanism to derive bearing capacity coefficients and employed the PSO technique to optimise these coefficients. The analysis considers the influence of slope inclination, soil cohesion, internal friction angles, unit weights and layer properties on bearing capacity coefficients. Key findings showed that seismic bearing capacity coefficients decrease significantly with increasing seismic coefficients and slope angles.

Vahidreza and Ravichandran [63] presented a robust climate-adaptive design optimisation framework for shallow foundations that incorporates site-specific geotechnical and hydrological uncertainties. They utilised a mathematical model based on principles of partially saturated soil mechanics to compute bearing capacity and settlement while employing the NSGA-II (non-dominated sorting genetic algorithm

II) for multi-objective optimisation. The proposed optimisation method predicted smaller foundation dimensions (width and depth), resulting in lower material usage and excavation costs. Consequently, the overall construction cost was decreased by nearly 19% compared to conventional design methods.

Nawaz et al. [49] presented a cost-based optimisation model for the design of isolated foundations in cohesive soils using the generalised reduced gradient (GRG) method, implemented through Microsoft Excel's Solver tool. The objective was to minimise the total construction cost by treating the footing's width, length and embedment depth as design variables while satisfying geotechnical safety constraints: a minimum FOS of 3 against bearing capacity failure (ULS) and a maximum allowable settlement of 25 mm (SLS). Applied to a design example in silty clay, the optimised model achieved a 44% cost reduction compared to a conventional design.

Waheed et al. [64] focused on developing a user-friendly, spreadsheet-based tool for optimising RC isolated footings using advanced metaheuristic algorithms, specifically the evolutionary algorithm (EA) and the GA. The tool aims to minimise the cost of RC isolated footings while adhering to safety and strength requirements outlined by ACI 318-14. They compared the optimised designs generated from the developed tool against the original designs and those optimised using the GRG algorithm, demonstrating the tool's efficiency and cost-saving potential. They allowed for analysis of four case studies, yielding cost reductions of up to 44%, and also explored the effects of geometric parameters, reinforcement ratios and material strengths on the optimised cost, providing insights into achieving cost-effective designs.

Kashani et al. [65] focused on optimising the design of combined footings using five swarm intelligence algorithms—PSO, APSO, MFO, ALO and WOA—under varying conditions. Two design scenarios were evaluated: one with moderate uniaxial loads and another with higher axial loads, considering both rectangular and trapezoidal geometries. The study revealed that PSO outperforms other optimisation methods, achieving a 17% reduction in final costs, particularly favouring trapezoidal footings. Key factors such as concrete compressive strength, soil friction angle and allowable settlement were analysed for their impact on costs, with optimal concrete strength identified between 40 and 50 MPa. Additionally, the research explored the effects of load variations and introduced alternative algorithms such as APSO and ALO to enhance solution efficiency. Overall, the findings underscore the importance of various design parameters in achieving cost-effective combined footing solutions.

Khajehzadeh et al. [66] presented an optimisation approach for spread foundations using the adaptive salp swarm algorithm (ASSA), which aims to minimise total cost while ensuring compliance with structural and geotechnical design criteria. This algorithm defines the objective function mathematically, incorporating unit costs for materials such as concrete and reinforcement, and outlines design variables and constraints related to bearing capacity, settlement and shear. Through case studies, the ASSA demonstrated superior performance in achieving optimal designs compared to traditional methods and other metaheuristic algorithms, highlighting its effectiveness in the economic design of spread footings.

Arabali et al. [67] presented a novel adaptive version of the tunicate swarm algorithm (ATSA) and evaluated its performance against standard algorithms and other metaheuristic methods using benchmark test functions. The study discussed the time complexity of the ATSA, the effectiveness of various OAs and the application of ATSA in the design of low-cost and low-CO_2 emission shallow foundations. Additionally, it highlighted the importance of accurately measuring parameters such as effective friction angle for optimal design outcomes.

Alzabeebee et al. [68] presented a novel symbolic predictive model, the multi-objective evolutionary polynomial regression computing (EPR-MOGA), for estimating the UBC of shallow foundations on cohesionless soils. Utilising a multi-objective GA, the model automatically derived an optimised mathematical relationship between key variables including foundation dimensions, soil properties and embedment depth. The approach integrated evolutionary computing techniques to explore a vast search space, resulting in simple, robust and accurate symbolic equations. The study conducted an extensive statistical assessment and compared the EPR-MOGA's performance with that of classical methods and other data-driven models, such as neural networks and regression techniques. Results demonstrated that EPR-MOGA outperforms traditional and existing soft computing models within the confines of the training data range, offering a reliable and interpretable tool for geotechnical engineers. The methodology emphasised the importance of data range limitations, highlighting that the model's predictions are most accurate within the dataset's variable bounds.

Waheed et al. [69] presented an optimisation-based approach for designing RC isolated stepped footings, aiming to enhance economic and eco-efficient performance compared to conventional pad footings. This approach applied a GA to find the optimal geometric and reinforcement factors, incorporating cost, embodied energy and carbon emissions as objective functions, revealing embodied energy minimisation as the most effective eco-efficient design strategy. Optimised stepped footings reduced material use and costs by adapting depth to load distribution, outperforming traditional pad footings in cost savings and material reduction.

Fattahi et al. [70] aimed to improve the prediction accuracy of shallow foundation settlement by applying two intelligent optimisation algorithms TLBO and HS to address limitations of traditional methods in handling non-linearities and uncertainties in geotechnical parameters. Using 189 data points with key inputs including footing width, pressure, standard penetration test (SPT) blow count, embedment ratio and geometry, models were trained and tested with normalisation and iterative refinement to minimise errors. Parameter SPT blow count was identified as the most influential factor affecting settlement predictions, emphasising its critical role in model accuracy. The models reduced costs and time by minimising extensive site testing, improving design accuracy and handling uncertainties effectively and were adaptable across various geotechnical projects.

Table 2 provides a summary of the key findings from the studies discussed above.

Table 2 showcases the various studies focused on cost optimising different types of footings. It highlights the diversity in types of datasets, design variables and optimisation methods employed. This reflects a trend towards utilising sophisticated optimisation methods, in particular GA and PSO, to reduce costs and improve performance based on specific input parameters, including geometric dimensions and soil properties. This body of research indicates that the geometric dimensions of the footing (length, width, thickness and embedment depth), along with the amount of reinforcement (area of steel, bar size/spacing), are the key factors that influence optimal design. Following these, soil properties, concrete strength and load characteristics are the next most important contributors.

In addition, these studies show that GA and PSO are the most commonly used optimisation techniques. This makes sense as both methods are well established and robust. They are especially effective for solving complex engineering design problems that involve multiple variables and constraints, which is often the case in foundation design optimisation.

Almost 85% of the studies used numerical or hybrid numerical-analytical methods, indicating a significant lack of experimental or field in situ data validation—a gap that should be taken into account in future research. Overall, emphasis is placed on using advanced computational techniques to improve foundation design efficiency and cost-effectiveness.

3.2 Cost Functions Available for Optimising the Design of Shallow Foundations

In today's research landscape, numerous scholars are exploring OAs as a method to determine the optimal design of shallow foundations. Almost all of them have tried to explicitly incorporate cost aspects to achieve economically optimised foundation designs while simultaneously not exceeding ULS and SLS requirements. Table 3 summarises the extant studies that utilise OAs to identify cost-effective designs.

As can be seen in Table 3, these authors have considered different components of the costs for the cost optimisation of the footing, with the three most common being $C_{form} + C_{rein} + C_{conc}$. This suggests that these are the governing factors for the costs. The construction cost estimated by each author varies due to differences in calculation methods, influenced by the unique considerations and standards of each country.

Jelušič and Žlender [56, 59] considered several unique costs in their optimisation footing, including C_{sta}, C_{fso}, C_{tran}, C_{cund}, C_{hins} and C_{tins}, and only Jelušič and Žlender [59] considered the C_{geo}. Considering these as part of the cost for the footing is not common, except for C_{tran} and C_{cund}. Different locations have varying methods for calculating C_{tran}. In most places, the cost of the transportation concrete is subject to the distance of the concrete supplier and the volume of the concrete. Although the quantity of the concrete can be determined based on the geometry, in actual

Table 2 Application of OAs for shallow foundation problems

Reference	Type of dataset	Type of footing	Design variables (inputs)	Most effective parameters for optimal design	Objectives (outputs)	Optimisation method
Nigdeli et al. [57]	Numerical	RC spread footing	Dimensions of the footing (base dimensions, height), eccentricity of the column and reinforcement details such as bar size and spacing	Geometric dimensions and reinforcement quantities	Cost of footing	Harmony search, flower pollination algorithm, teaching-learning-based optimisation algorithm
Gandomi and Kashani [58]	Numerical	RC shallow footing	Base soil friction angle, soil density, modulus of elasticity, Poisson's ratio, concrete compressive strength, force inclination, depth of footing	Depth of footing, concrete compressive strength, force inclination	Cost of shallow foundation	Accelerated particle swarm optimisation, firefly algorithm, whale optimisation algorithm, ant lion optimiser, Lévy flight krill herd, grey wolf optimiser, teaching-learning-based optimisation algorithm, moth-flame optimisation algorithm

(continued)

Table 2 (continued)

Reference	Type of dataset	Type of footing	Design variables (inputs)	Most effective parameters for optimal design	Objectives (outputs)	Optimisation method
Jelušič and Žlender [56]	Numerical-analytical	Pad footing	Pad footing length, width, thickness, diameter of bars, spacing between bars, concrete compressive strength	Load magnitudes, soil shear strength parameters, footing geometric dimensions, concrete strength, reinforcement characteristics	Structure production cost	Multiparametric mixed-integer non-linear programming optimisation
Jelušič and Žlender [59]	Numerical	Reinforced pad and strip foundations	Strip foundation width and thickness, pad depth, width of reinforced pad foundation, number of reinforcement layer, spacing between the geosynthetic reinforcement layers	Vertical permanent load, soil shear angle	Structure production cost	Multiparametric mixed-integer non-linear programming optimisation
Islam and Rokonuzzaman [48]	Numerical-analytical	Shallow isolated column footing	Footing width, length, thickness, embedment depth	Young's modulus and frictional angle of soil	Construction cost	Genetic algorithm
Azhim and Prakoso [60]	Numerical	Shallow foundation	Vertical load design, footing depth, width, length, area, soil properties (unit weight, friction angle, Young's modulus, Poisson's ratio)	Friction angle and Young's modulus	Construction cost	Solver add-in function in MS Excel

Reference	Type of dataset	Type of footing	Design variables (inputs)	Most effective parameters for optimal design	Objectives (outputs)	Optimisation method
Kashani et al. [6]	Numerical-analytical	RC shallow footing	Footing dimensions, reinforcement details, column location	Footing depth, concrete strength, load inclination	Cost of shallow foundation	Differential evolution, evolutionary strategy, biogeography-based optimisation
Chaudhuri and Maity [61]	Numerical-analytical	Isolated rectangular RC footing	Footing depth, width, steel reinforcement area	Plan area, depth, area of flexural reinforcement at moment critical sections, width of footing	Total footing cost	Binary-coded genetic algorithm, unified particle swarm optimisation
Debnath [71]	Numerical-analytical	Strip footing	Depth of embedment, cohesion ratio, slope inclination, seismic coefficient, thickness of top layer	Depth of embedment, cohesion ratio, slope inclination, seismic coefficient, thickness of top layer	Bearing capacity coefficients	Particle swarm optimisation
Vahidreza and Ravichandran [63]	Field (in situ), numerical, analytical	Shallow foundation	Foundation width and depth, geotechnical and hydrological parameters	Soil strength and deformation properties, matric suction, degree of saturation, rainfall and drought characteristics, groundwater depth	Minimise the standard deviation of the foundation's response (for robustness), construction cost	Non-dominated sorting genetic algorithm
Nawaz et al. [49]	Numerical	Isolated foundation	Embedment depth of foundation, width, length, thickness	Design load, Young's modulus, recompression index	Construction cost of foundation	Generalised reduced gradient

(continued)

Table 2 (continued)

Reference	Type of dataset	Type of footing	Design variables (inputs)	Most effective parameters for optimal design	Objectives (outputs)	Optimisation method
Waheed et al. [64]	Numerical-analytical	RC isolated footing	Footing thickness, length, width, area of steel, concrete compressive strength, steel yield strength	Reinforcement ratio, geometric dimensions (length, width, thickness), material strengths (concrete compressive, steel yield strength)	Total footing cost	Evolutionary algorithm, genetic algorithm
Kashani et al. [65]	Numerical	Combined footing	Six geometric parameters (e.g. distances from columns, widths, thickness and depth of the footing), seven steel reinforcement parameters (e.g. bar size/number)	Concrete compressive strength, soil friction angle, factor of safety against soil bearing capacity, elasticity modulus of soil	Total shallow footing cost	Particle swarm optimisation, accelerated PSO, ant lion optimiser, moth-flame optimiser, whale optimisation algorithm
Khajehzadeh et al. [66]	Numerical-analytical	Spread foundation	Geometric dimensions (length, width, thickness), required reinforcement	Geometric dimensions, especially length and width (shape)	Total foundation cost	Adaptive salp swarm algorithm
Arabali et al. [67]	Numerical	Shallow spread foundation	Geometric dimensions (length, width, thickness, depth of embedment), required reinforcement	Geometric dimensions (length, width, thickness, depth of embedment)	Total cost and CO_2 emissions	Adaptive tunicate swarm algorithm

Reference	Type of dataset	Type of footing	Design variables (inputs)	Most effective parameters for optimal design	Objectives (outputs)	Optimisation method
Alzabeebee et al. [68]	Experimental and field (in situ) data	Shallow foundation	Foundation width, embedment, dry unit weight of the soil, length to width ratio, angle of internal friction of the soil, ultimate bearing capacity	Foundation width, embedment, length, dry unit weight of soil	Estimating ultimate bearing capacity	Evolutionary polynomial regression with multi-objective genetic algorithm
Waheed et al. [69]	Numerical	RC isolated stepped footing	Foundation length, width, depth, area of steel	Depth of the footing at column face	Total cost, carbon emissions, embodied energy	Genetic algorithm
Fattahi et al. [70]	Field (in situ) data, experimental	Shallow foundation	Footing width, pressure exerted on footing, count of SPT blows, ratio of footing embedment, geometry of footing	SPT blow count	Prediction accuracy of settlement	Teaching-learning-based optimisation, harmony search

Table 3 List of construction cost models and associated OAs

Reference	Optimisation algorithm	Cost objective function
Nigdeli et al. [57]	Harmony search, flower pollination algorithm, teaching-learning-based optimisation algorithm	$C_{conc} + C_{rein}$ where C_{rein} is the reinforcing steel cost item (€/m) C_{conc} is the concrete cost item (€/m)
Gandomi and Kashani [58]	Accelerated particle swarm optimisation, firefly algorithm, Lévy flight krill herd, moth-flame optimisation algorithm, whale optimisation algorithm, ant lion optimisation, grey wolf optimiser, teaching-learning-based optimisation algorithm	$C_{exc} + C_{form} + C_{rein} + C_{conc} + C_{backfill}$ where C_{exc} is the ground excavation cost item (€/m) C_{form} is the formwork cost item (€/m) $C_{backfill}$ is the backfill cost item (€/m)
Jelušič and Žlender [56]	Multiparametric mixed-integer and non-linear programming	$C_{exc} + C_{sta} + C_{fso} + C_{tran} + C_{form} + C_{rein} + C_{cund} + C_{conc} + C_{hins} + C_{tins}$ where C_{sta} is the stabilisation of layer cost item (€/m) C_{fso} is the made ground cost item (€/m) C_{tran} is the transportation cost item (€/m) C_{cund} is the underlay concrete cost item (€/m) C_{hins} is the hydro-insulation cost item (€/m) C_{tins} is the thermos insulation cost item (€/m)
Jelušič and Žlender [59]	Multiparametric mixed-integer and non-linear programming	$C_{exc} + C_{sta} + C_{geo} + C_{fso} + C_{tran} + C_{form} + C_{rein} + C_{cund} + C_{conc} + C_{hins} + C_{tins}$ where C_{geo} is the geotextile cost item
Islam and Rokonuzzaman [48]	Genetic algorithm	$C_{conc} + C_{rein} + C_{exc} + C_{form} + C_{backfill}$
Azhim and Prakoso [60]	Solver add-in function in MS Excel	$C_{exc} + C_{form} + C_{rein} + C_{conc} + C_{backfill}$

Reference	Optimisation algorithm	Cost objective function
Kashani et al. [6]	Evolutionary algorithm	$C_{exc} + C_{form} + C_{rein} + C_{conc} + C_{backfill}$
Chaudhuri and Maity [61]	Genetic algorithm and unified particle swarm optimisation	$C_{form} + C_{rein} + C_{conc}$
Mahmoudabadi and Ravichandran [63]	Non-dominated sorting genetic algorithm	$C_{conc} + C_{rein} + C_{form} + C_{exc} + C_{backfill}$
Nawaz et al. [49]	Generalised reduced gradient method	$C_{exc} + C_{form} + C_{rein} + C_{conc} + C_{backfill}$
Waheed et al. [64]	Evolutionary algorithm, genetic algorithm	$C_{form} + C_{rein} + C_{conc}$
Kashani et al. [65]	Particle swarm optimisation, accelerated, ant lion optimiser, moth-flame optimiser, whale optimisation algorithm	$C_{exc} + C_{form} + C_{rein} + C_{conc} + C_{backfill}$
Khajehzadeh et al. [66]	Adaptive salp swarm algorithm	$C_{exc} + C_{form} + C_{rein} + C_{conc} + C_{backfill}$
Arabali et al. [67]	Adaptive tunicate swarm algorithm	$C_{conc} + C_{exc} + C_{backfill} + C_{form} + C_{rein}$
Waheed et al. [69]	Genetic algorithm	$C_{conc} + C_{rein}$

construction works, the footing foundation is not the only construction work carried out; other elements such as structural columns, ground beams and retaining walls are also involved. Therefore, it is not recommended to include this cost as it is quite subjective.

C_{cund}, also known as lean concrete, is used as a binding layer to create a smooth level surface and prevent contamination between the subsoil and the structural concrete with low compressive strength, usually at 5 MPa strength [72, 73]. Although this is a minor cost, it should be considered part of the overall cost.

Other than this, $C_{backfill}$ can be considered part of C_{exc}, because most pad footing foundation is adopted when the ground is in good condition. Thus, to simplify the calculation, $C_{backfill}$ can be considered under C_{exc}.

From the author's point of view, C_{exc}, C_{form}, C_{rein} and C_{conc} are the main and most common factors for the footing foundation works, as these are the components that govern most of the construction cost.

There are two types of reinforcements that can be found in C_{rein}: main reinforcement and shear reinforcement. Different countries have different computational methods for reinforcement; therefore, it is difficult to standardise the calculation method for the footing reinforcement calculation.

In general, for most common cases of footing optimisation, the cost objective function can be estimated via the following equation:

$$C_{exc} + C_{form} + C_{rein} + C_{conc} \tag{11}$$

Based on the above descriptions, it can be seen that many yet-to-be-explored optimisation approaches offer the potential to identify a suitable equation for the minimum cost of constructing footings. While optimisation techniques have been widely explored, PSO and Bayesian optimisation (BO) have emerged as promising approaches for optimising construction footing costs owing to their robust search capabilities and adaptability [57].

PSO, inspired by the swarm intelligence observed in nature, is an evolutionary algorithm that optimises a problem by iteratively improving candidate solutions [74, 75]. In structural engineering, PSO has been applied to optimise the design parameters of construction footings by reducing material costs while maintaining structural integrity. Several studies have demonstrated that PSO efficiently balances exploration and exploitation, making it well suited for cost minimisation in construction projects [22]. PSO operates by initialising a swarm of particles, each of which represents a possible solution. The particles navigate through the search space based on their own experience and those of their neighbours, updating their positions using velocity and position-update equations. The algorithm iterates until it converges on an optimal or near-optimal solution. In construction footing optimisation, PSO can be used to minimise material usage while ensuring adequate load-bearing capacity.

BO is a probabilistic model-based optimisation technique that uses prior information to guide the search for optimal solutions. Unlike traditional OAs, BO excels in problems where function evaluations are expensive, such as in construction design, where multiple simulations are required to validate a footing's performance. Recent research indicates that BO effectively reduces the number of required evaluations, leading to time and cost savings in optimising structural foundations [76]. BO employs a surrogate model, typically a Gaussian process (GP), to estimate the objective function's behaviour. By leveraging the posterior distribution of the function, BO intelligently selects new evaluation points that maximise information gain. In construction applications, BO can be used to optimise footing dimensions, reinforcement placement and material selection while reducing computational effort.

Both PSO and BO can be implemented in construction optimisation using the following approach:

1. Define the objective function (e.g. cost minimisation while maintaining structural integrity).
2. Establish design constraints (e.g. soil bearing capacity, safety factors and material availability).
3. Initialise the OA, and run simulations.
4. Evaluate results, and refine designs based on optimal parameters.

While PSO is effective in handling high-dimensional optimisation problems, BO excels when function evaluations are costly. The decision to use PSO or BO depends on the specific requirements of the project, such as available computational resources and the intricacies involved in designing a structural footing.

4 Discussion and Recommendations

There are many alternative optimisations that have yet to be considered, offering further possibilities for OAs. Therefore, it is strongly recommended to use additional optimisation techniques and hybrid models, before comparing to identify which is most suitable. The ongoing refinement of these models is expected to enhance their adaptability in different project scenarios. For the beginner, the author suggests that the optimisation process be initialised for a small-scale footing.

It would be valuable to increase experimental and field validation to bridge the gap between computational results and real-world performance and to investigate more complex types of foundations and loading conditions that go beyond basic rectangular footings. A general flowchart for the optimisation of a footing is illustrated in Fig. 2.

Based on this flowchart, the first stage collects the unit rate for the costs. After this, the footing design is required, where the formula for the computation is as shown in Table 4. The proposed summary of the table is recommended for a simplified and common design for the foundation. After identifying the influential parameters, the researcher must set the boundary for each of the costs. Optimisation is then used to identify the value that gives the lowest cost possible. After the completion of the OA analysis, it is required to check the input against the actual

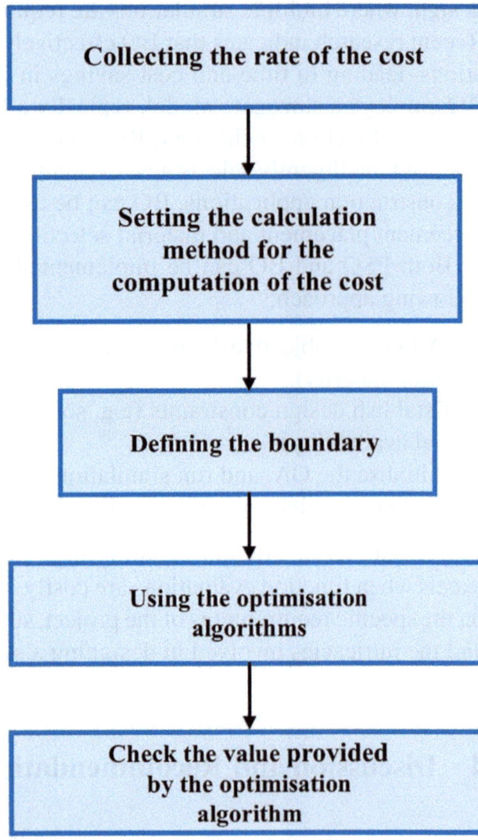

Fig. 2 General flowchart for the optimisation of footing design

Table 4 Cost items of the objective function

Cost items	Formula
Excavation for the footing works	$\left[\left(2\times\frac{1}{2}\times D_f\times(a+b)\times L\right)+\left(D_f\times B\times L\right)\right]\cdot c_{exc}$
Lean concrete	$(2a + B) \times L \cdot c_{lean}$
Formwork for the footing concrete	$[(2B + 2L) \times t \cdot c_{form}]$
Reinforcement bar of the footing	$\left[\left(\frac{d^2}{162}\right)\times\left(\frac{L}{s}+1\right)\times L+\left(\frac{d^2}{162}\right)\times\left(\frac{B}{s}+1\right)\times B\right]\times 1.05\times c_{rein}$
Concrete cost	$\left[\left(B\times L\times D_f\right)\times\frac{f_{ck,design}}{f_{20/25}}\right]\cdot c_{conc}$

Here, c_{exc} is the cost of excavation RM per m^3
c_{lean} is the cost of the lean concrete of 5 MPa RM per m^2
c_{form} is the formwork cost of RM per m^2
c_{rein} is the cost of the reinforcement of RM per kg
c_{conc} is the cost of the concrete of RM per m^3 for concrete grade 25 MPa
$\frac{d^2}{162}$ is the weight of the rebar reinforcement in kg

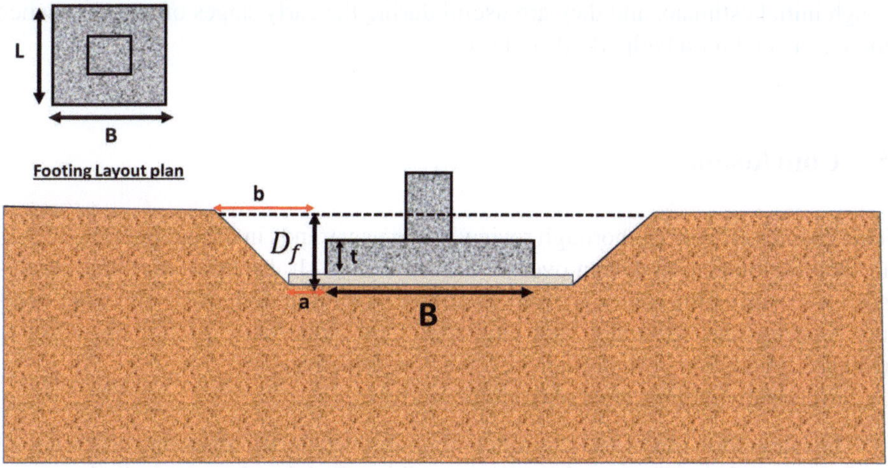

Fig. 3 Components of the footing design

calculation of the ULS and SLS to ensure that the output provided is reasonable and practical. Figure 3 illustrates the components of the footing.

To simplify the calculation, a constant value of 162 can be used, derived as follows:

Weight of the reinforcement bar based on the rebar density of 7850 kg m^{-3}:

$$\frac{\pi\left(\frac{d}{1000}\right)^2}{4} \times 1\,\text{m} \times 7850\,\text{kg}\,\text{m}^{-3} \tag{12}$$

$$\sim \frac{d^2}{162}\,\text{kg}$$

As for the reinforcement bar calculation, the main reinforcement bars are bottom reinforcement in both directions.

It is worth noting that the actual cost of concrete does not always scale linearly with strength (see the concrete cost in Table 4); for example, shifting from 20 to 25 MPa might imply a different cost increase than shifting from 25 to 30 MPa. Different concrete strengths often involve different mix designs (various proportions of cement, aggregates and admixtures). These changes in materials can affect the cost in ways that the simple ratio (i.e. the ratio of cement, fine and coarse aggregates, which is normally used for concrete mix design) does not capture. Concrete suppliers have their own pricing structures, which can be influenced by factors other than just the concrete strength (e.g. demand, availability of materials, transportation costs). High-strength concrete often uses special admixtures, which can significantly add to the cost. Using the formulas mentioned in the table can serve as a rough initial estimate, and they are useful during the early stages of a project when there is a need for a ballpark idea of costs.

5 Conclusion

This chapter provides a thorough review of the new trends in the application of OAs in shallow foundation design over the last few years. It discusses essential aspects such as ensuring the safety and serviceability of footing designs by checking bearing capacity and settlement and emphasises the importance of utilising OAs to enhance foundation design efficiency and cost-effectiveness.

Several key findings emerge from the study:

- The integration of OAs has significantly improved the accuracy and efficiency of footing designs. The transition from traditional trial-and-error methods to systematic optimisation techniques has not only sped up the design process but also enhanced overall structural integrity.
- Implementing OAs has shown significant cost savings, with reported reductions of up to 68% in some instances. This efficiency is accomplished while complying with essential geotechnical and structural standards, incorporating ULS and SLS.
- Geometric dimensions (length, width, thickness, embedment depth) consistently emerge as the most influential parameters affecting optimal footing design, directly affecting both cost and structural performance. Soil properties, particularly friction angle and Young's modulus, represent the second most important group of factors influencing optimisation outcomes.

Various OAs have been discussed, and a clear trend towards adaptive and hybrid algorithms for improved optimisation efficiency has been demonstrated. GA and PSO are the most prevalent optimisation methods, chosen for their robustness in handling complex engineering constraints.

AI Use Disclosure Statement The authors did not use any generative AI or AI-assisted technologies in the creation of this manuscript.

The authors used generative AI technologies only for proofreading. This includes minor corrections to grammar, spelling, and syntax. No AI was used for ideation, data analysis, or writing original content.

References

1. Craig RF, Knappett JA (2012) Craig's soil mechanics, 8th edn. Spon Press
2. Galli A, Mortara G (2021) Experiments and modelling of horizontal loading tests on small-scale foundations in sand. Geotech Lett 11:171–178. https://doi.org/10.1680/jgele.20.00098

3. He P, Fenton GA, Griffiths DV (2023) Load and resistance factor design versus reliability-based design of shallow foundations. Georisk 17:277–286. https://doi.org/10.1080/17499518.2022.2083179
4. Atkinson M (1993) Structural foundations manual for low-rise buildings. CRC Press
5. Magar J, Kudtarkar A, Pachpohe J, Nagargoje P (2020) Study and analysis of types of foundation and design construction. Int Res J Eng Technol 7:3301–3307. https://doi.org/10.5281/zenodo.3995061
6. Kashani AR, Gandomi M, Camp CV, Gandomi AH (2020) Optimum design of shallow foundation using evolutionary algorithms. Soft Comput 24:6809–6833. https://doi.org/10.1007/s00500-019-04316-5
7. Harandizadeh H, Jahed Armaghani D, Khari M (2021) A new development of ANFIS–GMDH optimized by PSO to predict pile bearing capacity based on experimental datasets. Eng Comput 37:685–700
8. Zhang W, Gu X, Tang L, Yin Y, Liu D, Zhang Y (2022) Application of machine learning, deep learning and optimization algorithms in geoengineering and geoscience: comprehensive review and future challenge. Gondwana Res 109:1–17. https://doi.org/10.1016/j.gr.2022.03.015
9. Boumezerane D (2022) Recent tendencies in the use of optimization techniques in geotechnics: a review. Geotechnics 2:114–132. https://doi.org/10.3390/geotechnics2010005
10. Koopialipoor M, Fahimifar A, Ghaleini EN, Momenzadeh M, Armaghani DJ (2020) Development of a new hybrid ANN for solving a geotechnical problem related to tunnel boring machine performance. Eng Comput 36:345–357
11. Hasanipanah M, Noorian-Bidgoli M, Jahed Armaghani D, Khamesi H (2016) Feasibility of PSO-ANN model for predicting surface settlement caused by tunneling. Eng Comput 32. https://doi.org/10.1007/s00366-016-0447-0
12. Mikaeil R, Shaffiee Haghshenas S, Sedaghati Z (2019) Geotechnical risk evaluation of tunneling projects using optimization techniques (case study: the second part of Emamzade Hashem tunnel). Nat Hazards 97:1099–1113. https://doi.org/10.1007/s11069-019-03688-z
13. Momeni E, Nazir R, Armaghani DJ, Maizir H (2014) Prediction of pile bearing capacity using a hybrid genetic algorithm-based ANN. Measurement 57:122–131
14. Moayedi H, Jahed Armaghani D (2018) Optimizing an ANN model with ICA for estimating bearing capacity of driven pile in cohesionless soil. Eng Comput 34:347–356
15. Jelušič P (2024) Enhancing sustainability of building foundations with efficient open-end pile optimization. Sustainability 16:1–15. https://doi.org/10.3390/su16166880
16. Haghshenas SS, Haghshenas SS, Geem ZW, Kim T-H, Mikaeil R, Pugliese L, Troncone A (2021) Application of harmony search algorithm to slope stability analysis. Land 10. https://doi.org/10.3390/land10111250
17. Koopialipoor M, Armaghani DJ, Hedayat A, Marto A, Gordan B (2018) Applying various hybrid intelligent systems to evaluate and predict slope stability under static and dynamic conditions. Soft Comput 23:5913–5929. https://doi.org/10.1007/s00500-018-3253-3
18. Mishra M, Gunturi VR, Miranda TFDS (2019) Slope stability analysis using recent metaheuristic techniques: a comprehensive survey. SN Appl Sci 1:1–17. https://doi.org/10.1007/s42452-019-1707-6
19. Lalicata LM, Bressan A, Pittaluga S, Tamellini L, Gallipoli D (2025) An efficient slope stability algorithm with physically consistent parametrisation of slip surfaces. Int J Civ Eng 23:671–682. https://doi.org/10.1007/s40999-024-01053-1
20. Mishra M, Ramana GV, Maity D (2020) Multiverse optimisation algorithm for capturing the critical slip surface in slope stability analysis. Geotech Geol Eng 38:459–474. https://doi.org/10.1007/s10706-019-01037-2
21. Mafi R, Javankhoshdel S, Cami B, Jamshidi Chenari R, Gandomi AH (2021) Surface altering optimisation in slope stability analysis with non-circular failure for random limit equilibrium method. Georisk 15:260–286. https://doi.org/10.1080/17499518.2020.1771739

22. Kashani AR, Chiong R, Mirjalili S, Gandomi AH (2021) Particle swarm optimization variants for solving geotechnical problems: review and comparative analysis. Arch Comput Methods Eng 28:1871–1927. https://doi.org/10.1007/s11831-020-09442-0
23. Ebrahim Noroozi Ghaleini MK (2018) A combination of artificial bee colony and neural network for approximating the safety factor of retaining walls. Eng Comput 35:647–658
24. Koopialipoor M, Murlidhar BR, Hedayat A, Armaghani DJ, Gordan B, Mohamad ET (2019) The use of new intelligent techniques in designing retaining walls. Eng Comput 36:283–294
25. Kashani AR, Gandomi AH, Azizi K, Camp CV (2022) Multi-objective optimization of reinforced concrete cantilever retaining wall: a comparative study. Struct Multidiscip Optim 65:1–19. https://doi.org/10.1007/s00158-022-03318-6
26. Sharma S, Saha AK, Lohar G (2022) Optimization of weight and cost of cantilever retaining wall by a hybrid metaheuristic algorithm. Eng Comput 38:2897–2923. https://doi.org/10.1007/s00366-021-01294-x
27. Andrab SG, Hekmat A, Bin YZ (2017) A review: Evolutionary computations (GA and PSO) in geotechnical engineering. Comput Water Energy Environ Eng 06:154–179. https://doi.org/10.4236/cweee.2017.62012
28. Hajihassani M, Jahed Armaghani D, Kalatehjari R (2018) Applications of particle swarm optimization in geotechnical engineering: a comprehensive review. Geotech Geol Eng 36:705–722. https://doi.org/10.1007/s10706-017-0356-z
29. Hashemi SM, Rahmani I (2018) Numerical comparison of the performance of genetic algorithm and particle swarm optimization in excavations. Civ Eng J 4:2186–2196. https://doi.org/10.28991/cej-03091149
30. Yin ZY, Jin YF, Shen JS, Hicher PY (2018) Optimization techniques for identifying soil parameters in geotechnical engineering: comparative study and enhancement. Int J Numer Anal Methods Geomech 42:70–94. https://doi.org/10.1002/nag.2714
31. Sadrossadat E, Basarir H, Luo G, Karrech A, Durham R, Fourie A, Elchalakani M (2020) Multi-objective mixture design of cemented paste backfill using particle swarm optimisation algorithm. Miner Eng 153:1–7. https://doi.org/10.1016/j.mineng.2020.106385
32. Jellali B, Frikha W (2017) Constrained particle swarm optimization algorithm applied to slope stability. Int J Geomech 17:1–6. https://doi.org/10.1061/(asce)gm.1943-5622.0001019
33. Gordan B, Koopialipoor M, Clementking A, Tootoonchi H, Mohamad ET (2018) Estimating and optimizing safety factors of retaining wall through neural network and bee colony techniques. Eng Comput 35:945–954
34. Mikaeil R, Beigmohammadi M, Bakhtavar E, Haghshenas SS (2019) Assessment of risks of tunneling project in Iran using artificial bee colony algorithm. SN Appl Sci 1:1–9. https://doi.org/10.1007/s42452-019-1749-9
35. Samui P, Hoang ND, Nhu VH, Nguyen ML, Ngo PTT, Bui DT (2019) A new approach of hybrid bee colony optimized neural computing to estimate the soil compression coefficient for a housing construction project. Appl Sci 9:1–18. https://doi.org/10.3390/app9224912
36. Koopialipoor M, Ghaleini EN, Haghighi M, Kanagarajan S, Maarefvand P, Mohamad ET (2019) Overbreak prediction and optimization in tunnel using neural network and bee colony techniques. Eng Comput 35:1191–1202. https://doi.org/10.1007/s00366-018-0658-7
37. Moayedi H, Gör M, Kok Foong L, Bahiraei M (2021) Imperialist competitive algorithm hybridized with multilayer perceptron to predict the load-settlement of square footing on layered soils. Meas J Int Meas Confed 172:1–14. https://doi.org/10.1016/j.measurement.2020.108837
38. Tashayo B, Behzadafshar K, Soltani Tehrani M, Afkhami Banayem H, Hashemi MH, Taghavi Nezhad SS (2019) Feasibility of imperialist competitive algorithm to predict the surface settlement induced by tunneling. Eng Comput 35:917–923. https://doi.org/10.1007/s00366-018-0641-3
39. Wu M, Cai Q, Shang T (2019) Assessing the suitability of imperialist competitive algorithm for the predicting aims: an engineering case. Eng Comput 35:627–636. https://doi.org/10.1007/s00366-018-0621-7

40. Crisp MP, Jaksa M, Kuo Y (2020) Optimal testing locations in geotechnical site investigations through the application of a genetic algorithm. Geosciences 10:1–18. https://doi.org/10.3390/geosciences10070265
41. Simpson AR, Priest SD (1993) The application of genetic algorithms to optimisation problems in geotechnics. Comput Geotech 15:1–19. https://doi.org/10.1016/0266-352X(93)90014-X
42. He BH, Du XL, Bai MZ, Yang JW, Ma D (2024) Inverse analysis of geotechnical parameters using an improved version of non-dominated sorting genetic algorithm II. Comput Geotech 171:1–10. https://doi.org/10.1016/j.compgeo.2024.106416
43. Yin Z y, Jin Y f, Liu Z q (2020) Practice of artificial intelligence in geotechnical engineering. J Zhejiang Univ Sci A 21:407–411. https://doi.org/10.1631/jzus.A20AIGE1
44. Terzaghi K (1943) Theoretical soil mechanics. Wiley, New York, p 314
45. Meyerhof GG (1963) Some recent research on the bearing capacity of foundations. Can Geotech J 1:16–26. https://doi.org/10.1139/t63-003
46. Pantelidis L, Griffiths DV (2015) Footing on the crest of slope: slope stability or bearing capacity? Eng Geol Soc Territ 2:1231–1234. https://doi.org/10.1007/978-3-319-09057-3_215
47. Moghaddam PT, Dehkordi PF, Ghazavi M (2023) Reliability analysis-based safety factor for stability of footings on frictional soils. Geomech Eng 33:543–552. https://doi.org/10.12989/gae.2023.33.6.543
48. Islam MS, Rokonuzzaman M (2018) Optimized design of foundations: an application of genetic algorithms. Aust J Civ Eng 16:46–52. https://doi.org/10.1080/14488353.2018.1445437
49. Nawaz MN, Ali AS, Jaffar STA, Jafri TH, Oh TM, Abdallah M, Karam S, Azab M (2022) Cost-based optimization of isolated footing in cohesive soils using generalized reduced gradient method. Buildings 12:1–19. https://doi.org/10.3390/buildings12101646
50. Rodríguez CA, Rodríguez-Pérez ÁM, López R, Hernández-Torres JA, Caparrós-Mancera JJ (2023) A finite element method integrated with Terzaghi's principle to estimate settlement of a building due to tunnel construction. Buildings 13:1–23. https://doi.org/10.3390/buildings13051343
51. Ekbote AG, Nainegali L (2021) Finite element analysis of two nearby interfering asymmetric footings embedded in cohesionless foundation medium. Geomech Geoeng 16:263–276. https://doi.org/10.1080/17486025.2019.1664776
52. Hamderi M (2018) Comprehensive group pile settlement formula based on 3D finite element analyses. Soils Found 58:1–15. https://doi.org/10.1016/j.sandf.2017.11.012
53. Ekbote AG, Nainegali L (2019) Interference of two closely spaced footings embedded in unreinforced and reinforced soil medium: a finite element approach using ABAQUS. Arab J Geosci 12:1–21. https://doi.org/10.1007/s12517-019-4868-0
54. Poulos HG, Davis EH (1991) Elastic solutions for soil and rock mechanics. University of Sydney
55. (CEN) TEC for S (2004) Eurocode 2: design of concrete structures. European Committee for Standardization (CEN)
56. Jelušič P, Žlender B (2018) Optimal design of pad footing based on MINLP optimization. Soils Found 58:277–289. https://doi.org/10.1016/j.sandf.2018.02.002
57. Nigdeli SM, Bekdaş G, Yang XS (2018) Metaheuristic optimization of reinforced concrete footings. KSCE J Civ Eng 22:4555–4563. https://doi.org/10.1007/s12205-018-2010-6
58. Gandomi AH, Kashani AR (2018) Construction cost minimization of shallow foundation using recent swarm intelligence techniques. IEEE Trans Ind Informatics 14:1099–1106. https://doi.org/10.1109/TII.2017.2776132
59. Jelušič P, Žlender B (2018) Optimal design of reinforced pad foundation and strip foundation. Int J Geomech 18:1–11. https://doi.org/10.1061/(asce)gm.1943-5622.0001258
60. Azhim A, Prakoso WA (2019) Construction cost optimization of shallow foundation for sand soil in Indonesia. MATEC Web Conf 270:05005. https://doi.org/10.1051/matecconf/201927005005

61. Chaudhuri P, Maity D (2020) Cost optimization of rectangular RC footing using GA and UPSO. Soft Comput 24:709–721. https://doi.org/10.1007/s00500-019-04437-x
62. Debnath L (2021) Seismic bearing capacity of shallow strip footing embedded in slope resting on two-layered soil. Stud Geotech Mech 43:285–306. https://doi.org/10.2478/sgem-2021-0021
63. Vahidreza M, Ravichandran N (2021) Robust climate-adaptive shallow foundation design optimization subjected to hydrological loads. IFCEE 2021:372–383. https://doi.org/10.1061/9780784483428.038
64. Waheed J, Azam R, Riaz MR, Shakeel M, Mohamed A, Ali E (2022) Metaheuristic-based practical tool for optimal design of reinforced concrete isolated footings: development and application for parametric investigation. Buildings 12:1–21. https://doi.org/10.3390/buildings12040471
65. Kashani AR, Camp CV, Akhani M, Ebrahimi S (2022) Optimum design of combined footings using swarm intelligence-based algorithms. Adv Eng Softw 169:1–18. https://doi.org/10.1016/j.advengsoft.2022.103140
66. Khajehzadeh M, Iraji A, Majdi A, Keawsawasvong S, Nehdi ML (2022) Adaptive salp swarm algorithm for optimization of geotechnical structures. Appl Sci 12:1–23. https://doi.org/10.3390/app12136749
67. Arabali A, Khajehzadeh M, Keawsawasvong S, Mohammed AH, Khan B (2022) An adaptive tunicate swarm algorithm for optimization of shallow foundation. IEEE Access 10:39204–39219. https://doi.org/10.1109/ACCESS.2022.3164734
68. Alzabeebee S, Alshkane YMA, Keawsawasvong S (2023) New model to predict bearing capacity of shallow foundations resting on cohesionless soil. Geotech Geol Eng 41:3531–3547. https://doi.org/10.1007/s10706-023-02472-y
69. Waheed J, Azam R, Riaz MR, Shakeel M (2025) Optimization-based innovative approach for economical design of reinforced concrete isolated footings. Innov Infrastruct Solut 10:1–13. https://doi.org/10.1007/s41062-024-01823-9
70. Fattahi H, Ghaedi H, Armaghani DJ (2025) Improving shallow foundation settlement prediction through intelligent optimization techniques. Comput Model Eng Sci 143:747–766. https://doi.org/10.32604/cmes.2025.062390
71. Debnath L (2025) Seismic bearing capacity of strip footing on two-layered soil: analytical and numerical approach. Int J Geotech Eng 19:700–721. https://doi.org/10.1080/19386362.2025.2530989
72. Venkatarama Reddy BV, Chander A (2009) Fly-ash-blended lean concrete mixes for concrete blocks. Proc Inst Civ Eng Constr Mater 162:59–66. https://doi.org/10.1680/coma.2009.162.2.59
73. Mishra T, French P, Sakkal Z (2013) Engineering a better road: use of two-way prestressed, precast concrete pavement for rapid rehabilitations. PCI J 58:129–141. https://doi.org/10.15554/pcij.01012013.129.141
74. Eberhart R, Kennedy J (1995) Particle swarm optimization. In: Proceedings of ICNN'95—International conference on neural networks, vol 4, pp 1942–1948
75. Wang D, Tan D, Liu L (2018) Particle swarm optimization algorithm: an overview. Soft Comput 22:387–408. https://doi.org/10.1007/s00500-016-2474-6
76. Malkomes G, Garnett R (2018) Automating Bayesian optimization with Bayesian optimization. Adv Neural Inf Process Syst 2018:5984–5994

Generalizable ML Solutions for Predicting Equivalent Viscous Damping Ratio of Diagonal RC Coupling Beams

Bilal Younis and Hao Wu

Abstract Diagonally reinforced concrete coupling beams (DRCBs) play an important role in coupled shear-wall systems for high-rise buildings, acting as primary energy dissipators under seismic loading. In current practice, their seismic design depends on adequate shear capacity and detailing—especially diagonal reinforcement and end confinement—to ensure sufficient ductility (rotation capacity) and energy dissipation. However, in recent years, the equivalent viscous damping ratio (EVDR) of coupling beams has not been systematically studied, and its use in simplified assessment remains under-documented. With this motivation, this study compiles a database of 125 DRCBs from experimental programs conducted between 1980 and 2025. We first performed a sensitivity analysis by deriving feature weights over 100 random trials and then evaluated performance using feature subsets ordered from highest to lowest weight. The analysis indicates that $\left(f_c^{'}, f_{yd}, f_{yl}, p_d, p_v, p_l\right)$ and aspect ratio (AR) are the most influential predictors of EVDR. Two machine learning (ML) solutions were developed—a single-learner multilayer perceptron (MLP) and an ensemble extreme gradient boosting (XGBoost) model. XGBoost achieved superior accuracy ($R^2 = 0.98$, $RMSE = 0.008$, $MAE = 0.005$) and exhibited minimal bias ($ME = 0.001$ and $\sigma_{ME} = 0.008$; overall balanced grade (OBG) = 0.007). Results show that XGBoost offers the best robustness and generalization for EVDR prediction. A practical solution (PS) was also derived using a minimal MLP structure to provide an explicit equation for practitioners. For validation, a blind test using two specimens from the literature was conducted; predictions from both trained models closely matched the experimental EVDR, demonstrating the generalization and reliability of the developed solutions for real-world applications.

Keywords RC coupling beams · Diagonally reinforced coupling beams · Machine learning · Equivalent viscous damping · Coupled shear-wall system

---

B. Younis · H. Wu (✉)
State Key Laboratory of Disaster Reduction in Civil Engineering, Tongji University, Shanghai, China
e-mail: bilal144@tongji.edu.cn; 2009wuhao_tom@tongji.edu.cn

M. Mirrashid et al. (eds.), *Optimization and Adaptive Control Algorithms*, Emerging Trends in Mechatronics, https://doi.org/10.1007/978-981-95-8387-4_11

1 Introduction

In high-rise buildings, coupling beams connect adjacent wall piers across openings (e.g., doors and windows) in coupled shear-wall systems [1]. By linking the walls, these beams transfer shear and couple the flexural actions of the piers, so that the walls act together rather than independently under seismic action. During seismic events, coupling beams act as the primary ductile mechanism, dissipating energy and transferring shear and moment within the coupled-wall system [2]. Because coupling beams are typically short to moderate in span and carry high shear demands, special reinforcement detailing and confinement are used to achieve a large rotation capacity and a stable cyclic response. Their strength and ductility directly influence the coupling ratio and the overall drift of the structure [3].

In early practice, only conventional (horizontal) reinforcement was used in coupling beams; however, experimental studies on conventionally reinforced beams showed that their energy dissipation capacity and ductility were insufficient [4, 5]. To address this, Paulay and Binney [6] introduced an X-shaped diagonal bar layout ("diagonal reinforcement"). Diagonally reinforced coupling beams (DRCBs) dissipate more energy and develop higher shear strength than conventionally reinforced beams, and they can achieve high ductility when diagonal bars and ties are adequately detailed [7–9]. High-fidelity nonlinear finite-element (FE) models can capture the complex cyclic behavior of RC coupling beams, but simplified methods remain attractive for efficiency [10]. Global peak demands for coupled-wall buildings can be estimated using inelastic spectra or equivalent-linear SDOF representations, provided that the effective stiffness and damping (or modification factors) are calibrated to the beam detailing and demand level. In particular, detailing (e.g., diagonal reinforcement), low span-to-depth ratios, and shear-flexure interaction induce pinching and degradation that must inform these modifiers, while axial restraint and slab participation shift the effective properties and the coupling ratio [11]. The concept of equivalent viscous damping dates back to Jacobsen (1930) [12] and has been studied extensively by subsequent researchers [13–15].

In recent years, several researchers have focused on improving the seismic response of DRCBs [16, 17]. Amiri et al. [18] evaluated the influence of key detailing parameters on DRCBs. Data-driven approaches using ML algorithms have also been widely applied to coupling-beam problems; for example, Zhao et al. [19] investigated the shear-strength capacity of plate-reinforced composite (PRC) coupling beams using ML; Hu et al. [20] predicted the peak load and peak beam rotation of coupling beams with XGBoost; Yaghoubi et al. [21] estimated the equivalent viscous damping for RC shear walls via ML; and Li et al. [22] developed surrogate ML models (random forest, CatBoost) trained on nonlinear time-history results to rapidly predict inter-story drift and floor acceleration in coupled-wall systems with hybrid coupling beams. Jafari et al. [23] proposed a novel MLP-based prediction model for the shear capacity of FRP-reinforced concrete (FRP-RC) beams. Collectively, these studies highlight both detailing-based strategies and machine learning-based prediction as complementary avenues for enhancing the seismic performance assessment of DRCBs.

Based on the literature review, the EVDR of RC coupling beams has not been systematically studied. To address this gap, a comprehensive database of DRCBs from worldwide experimental programs was established. A brief input-feature sensitivity analysis was performed to identify parameters most strongly influencing EVDR by deriving feature weights and evaluating the marginal contribution of individual features and ordered subsets. Using the dataset, single-learning and ensemble-learning ML models were developed to predict EVDR across θ levels. In addition to this, a PS was developed in the form of an easy-to-use mathematical equation, using a minimalistic MLP architecture with just one neuron in the hidden layer. This enables users to estimate EVDR using the required input features. Given the high performance of the predictive models, blind validation on unseen samples was conducted, with results demonstrating robust generalization and reliability for real-world scenarios. This independent analysis established a benchmark for the predictive power and reliability of the proposed solutions.

2 Database Establishment

This study established a comprehensive experimental database of 125 DRCBs from 33 peer-reviewed experimental studies [4, 8, 16, 24–53]. The database focuses solely on reinforced-concrete coupling beams detailed with diagonal reinforcement and all specimens tested under reversed quasi-static cyclic loading, ensuring consistency in the hysteretic response measures required for seismic calculation. It has a wide range of design and test features—including loading protocols with repeated θ levels to enable analysis of features' influences on inelastic response and equivalent viscous damping. A schematic of a coupled shear-wall system and a representative DRCB is provided in Fig. 1 to contextualize the member's role in force transfer and energy dissipation between adjacent wall piers during seismic excitation.

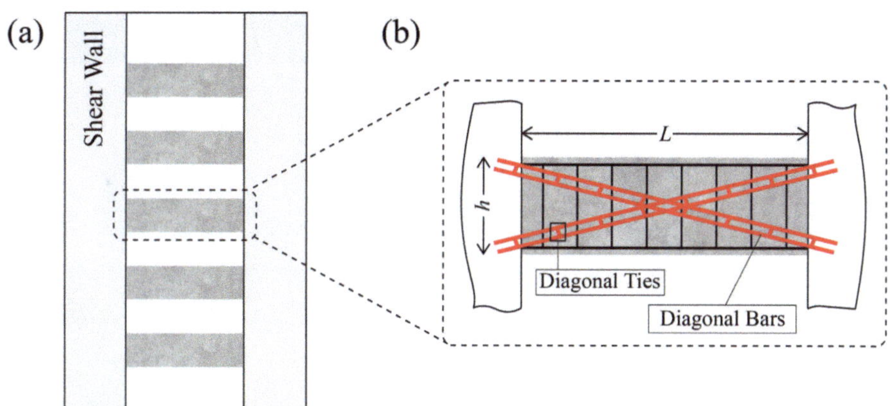

Fig. 1 Schematic illustration of (**a**) coupled shear-wall system and (**b**) diagonally reinforced coupling beam

3 Selection of Input Features and Output Variable

The preliminary data processing focused on analyzing the distribution of key features (independent parameters) in the dataset to assess their variability and ensure their suitability for further analysis. To analyze and understand the response of DRCBs, 14 design features as model input features including θ were selected for solution development. Selected features were kept based on their physical meaning to the coupling beams. Detailed descriptions of the data inside different scenarios were presented in the form of histograms and presented in Fig. 2.

Each plot provides brief insights into the frequency distribution of these features. The x-axis of each subplot shows minimum to maximum range of the feature. Due to the investigation of the seismic response of coupling beams, EVDRs were kept as model output. In the collected database, 33.6% of specimens failed in shear (S), 22.4% by diagonal bar fracture (DF), 20.0% by diagonal bar buckling (DB), 15.2% in flexure (F), and 8.8% in flexure-shear (FS). Data preprocessing began with digitizing the collected hysteresis loops for each specimen by correctly aligning the x- and y-axes. Many studies used different measurement units according to their own standards, but all extracted measurements were converted into consistent units (e.g., millimeters for dimensions, MPa for material strengths, percentage for reinforcement ratios). This study estimated the response of whole hysteric behavior at each θ level of the specimens. θ has a mean value of 2.89 ($\sigma = 2.02$ %); most cycles fall around 0.5–4% of θ, with a long sparse tail toward 10–12%. This reflects typical laboratory protocols focused on service to moderate damage states. EVDR unimodal is centered around $\mu = 0.18$ ($\sigma = 0.06$), with most values between 0.10 and 0.25. Lower EVDR reflects limited inelasticity or pinched hysteresis caused by slip

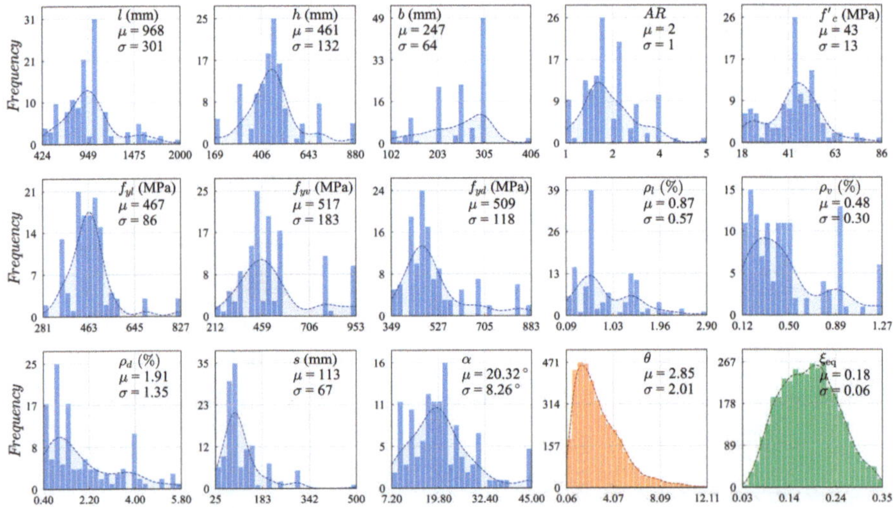

Fig. 2 Frequency distribution of the input features and output in the database

Table 1 Design properties of the specimens in the compiled database

Input features	μ	σ	Skew	$CoV(\%)$	Min	Max
Length (l) (mm)	968	301	0.89	31.20	424	2000
Height (h) (mm)	461	132	0.82	28.85	169	880
Width (b) (mm)	247	64	−0.55	26.20	102	406
Aspect ratio (AR) (l/h)	2	1	0.80	32.14	1	5
Concrete compressive strength $\left(f_c'\right)(MPa)$	43	13	0.11	29.78	18	86
Yield strength of longitudinal reinforcement (f_{yl}) (MPa)	467	86	1.84	18.56	281	827
Yield strength of stirrups reinforcement (f_{yv}) (MPa)	517	183	1.12	35.50	212	953
Yield strength of diagonal reinforcement (f_{yd}) (MPa)	509	118	1.56	23.28	349	883
Longitudinal reinforcement ratio (ρ_l) (%)	0.87	0.57	0.97	65.01	0.09	2.90
Stirrup reinforcement ratio (ρ_v) (%)	0.48	0.30	1.07	63.52	0.12	1.27
Diagonal reinforcement ratio (ρ_d) (%)	1.91	1.35	1.02	71.03	0.40	5.80
Stirrup spacing (s) (mm)	113	67	2.52	59.69	25	500
Angle of diagonal bars (α) (°)	20.32	8.26	1.05	40.78	7.20	45
Drift ratio (θ) (%)	2.85	2.01	1.29	70.72	0.06	12.11
Output feature						
Equivalent viscous damping ratio (ξ_{eq})	0.18	0.06	0.20	35.62	0.03	0.35

and inadequate confinement. Higher values are associated with stable yielding of the diagonal system and well-confined nodes that produce broad, non-pinched loops at moderate–large θ. However, extreme degradation can eventually reduce EVDR despite large θ. Calculation procedure of EVDR is given in the following section. A brief statistical description of the design features and output is provided in Table 1.

4 Calculation Procedure of Equivalent Viscous Damping Ratio

This section describes the procedure used to compute the EVDR consistently across all specimens. Starting from the recorded force–displacement ($F - \Delta$) hysteresis, each loading amplitude is segmented into fully closed cycles using zero crossings to ensure complete loop coverage. The EVDR is evaluated using Jacobsen's [12, 54] steady-state resonance formulation, as given in Eq. (1):

$$\xi_{eq} = \frac{1}{4\pi}\frac{E_D}{E_{so}} \tag{1}$$

Here, ξ_{eq} is the equivalent viscous damping ratio, E_D is the cyclic energy dissipated by the hysteresis loop, and E_{so} is the elastic reference energy for the same cycle. The elastic energy is computed from the peak ordinates of that cycle using Eq. (2):

$$E_{so} = \frac{1}{2} k u_0^2 = \frac{1}{2} f_0 \, u_0 \tag{2}$$

where k is the secant (equivalent) stiffness to the peak, u_0 is the maximum lateral displacement, and f_0 is the corresponding lateral force at u_0. After computing the cycle-level pairs (θ, ξ_{eq}) for each specimen, the θ support is harmonized via specimen-wise linear interpolation between the observed minimum and maximum θ. All original measurements were preserved (and flagged), and there was no extrapolation beyond each specimen's range. As a result, the learning algorithm can compare response at comparable θ levels. This approach captures the global (θ, ξ_{eq}) trend more precisely (including plateaus at larger θ), and estimate interaction with geometry, materials, and reinforcement detailing with improved stability.

5 Feature Sensitivity Analysis

Sensitivity analysis was conducted to determine the features that most influence the EVDR. Neighborhood component analysis (NCA), a well-known and widely adopted method for getting feature weights for structural parameters, was employed to quantify the relative importance of each input feature [55]. In this framework, NCA assigned a weight factor to each feature, enabling identification of those with the greatest influence on the model output. Figure 3 presents the outcomes as a heatmap: the x-axis indicates the input features, and the y-axis shows the number of trials conducted. To ensure robustness, 100 random trials were performed, and the mean weight for each feature across these trials is then normalized to the range of 0–1. Each column corresponds to the weight of a specific feature within a given trial, providing a clear, trial-by-trial view of variability and stability. Color shading indicates the importance of the feature: lighter shades indicate higher weights for influential features, whereas darker shades indicate lower weights with minimal impact on the EVDR. This visual representation makes it straightforward to see which features consistently contribute more and which contribute less to the EVDR. Overall, the NCA-based sensitivity analysis reveals an interpretable ranking of feature importance.

After obtaining feature weights using NCA, sequential forward feature selection was applied: subsets were created by adding features from the highest to the lowest weight. Figure 4 presents the mean values of R^2 and *RMSE* over 500 randomized trials for each subset to ensure a robust assessment of predictive performance. Consistent with the heatmap in Fig. 3, θ received the highest weight unsurprisingly, as increases in θ directly affect EVDR. With θ alone, the model achieved mean performance of (R^2, *RMSE*) = (0.43, 0.14). As higher-ranked features were appended $f_c', \rho_d, f_{yd}, f_{yl}, \alpha°, s, \rho_v, \rho_l, AR, l, and\, b$, R^2 increased monotonically while *RMSE* decreased. By the time subset reached b, performance reached R^2 = 0.958 and *RMSE* = 0.013, indicating incremental gains in explanatory power and predictive

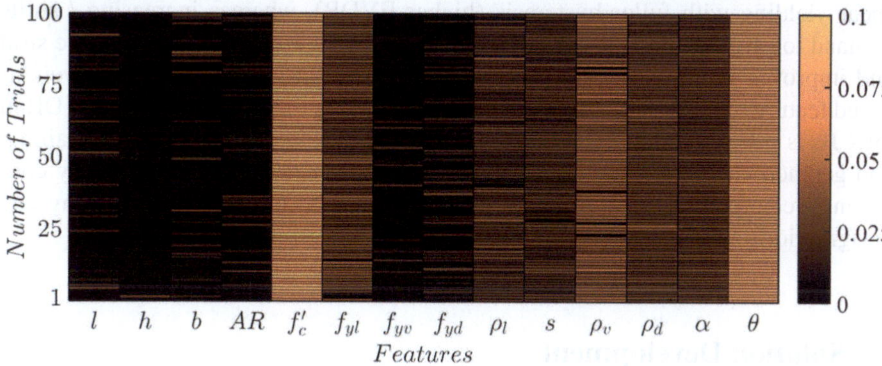

Fig. 3 Feature weights computed using NCA method over 100 random trials

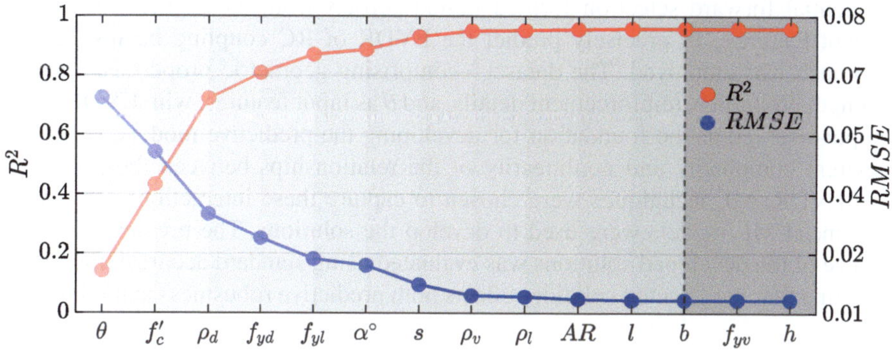

Fig. 4 Predictive performance of each subset: Mean values of R^2 and *RMSE* from the analysis over 500 trials

accuracy. Once the lower-weight features f_{yv} and h were included, both metrics effectively plateaued, showing no material improvement. This behavior corroborates the NCA ranking: the top 12 features drive EVDR prediction, whereas the remaining variables add minimal information.

It is necessary to understand that the obtained weight-based ranking and sensitivity of the features are transparent and align with the mechanics of the problem according to previously conducted experimental studies. Higher $f_c^{'}$ increases shear capacity and stiffness of the concrete strut and improves confinement effectiveness under cyclic compression; it also delays the crushing and helps to stabilize response, raising energy dissipation capacity, especially in short shear-dominant beams [52, 56]. EVDR in DRCBs is governed by yielding of the diagonal reinforcement; ρ_d *and* f_{yd} set the force level and plastic strain range that diagonal can sustain—higher grades raise shear capacity but demand tighter detailing to preserve cyclic ductility and EVDR [4, 6, 57]. Geometric properties of coupling beams $AR = l/h$, span l, and width b regulate the shear-flexure mechanism, strut area, and confinement effectiveness. Lower *AR* (shorter l for a given h) promotes diagonal-tension/

shear yielding with fuller hysteresis (higher EVDR), whereas increasing l shifts demand toward flexure and end rotations, and a larger b enlarges the concrete strut and improves stability, helping sustain EVDR [58]. In general, the NCA weight-based feature ranking and sequential forward selection jointly suggest that EVDR in DRCBs is governed chiefly by concrete strength, diagonal reinforcement detailing, and geometrical properties. These results support a parsimonious, physically consistent predictor set for EVDR and provide actionable guidance for detailing and design prioritization.

6 Solution Development

Feature sensitivity analysis was first performed using the NCA method; then sequential forward selection formed nested subsets from the highest- to lowest-weight features. To precisely predict the EVDR of RC coupling beams, an ML approach was employed. The dataset—comprising geometric properties, material strength properties, reinforcement details, and θ as input features, with EVDR as the output—served as the foundation for developing the predictive models. Given the inherent complexity and nonlinearity of the relationships between these features and EVDR, ML techniques were chosen to capture these interactions effectively. Advanced ML models were used to develop the solutions. The prediction performance of the developed solutions was evaluated using standard accuracy measures, ensuring that the selected solution reflects both predictive robustness and reliability.

6.1 Employed Algorithms

In this study, two algorithms were employed from two families: a single-learning algorithm, the multilayer perceptron (MLP), and an ensemble-learning algorithm, extreme gradient boosting (XGBoost). A practical solution (PS)—using a minimal MLP structure—was derived with a single hidden neuron to yield a closed-form, user-friendly equation for practical use (details in the following section). To enhance performance beyond the PS, a standard MLP was utilized. The MLP is a feed-forward network comprising an input layer, one or more hidden layers, and a linear output layer for regression [59]. Each hidden layer applies an affine transform followed by a nonlinear activation (e.g., ReLU, tanh, and logistic-sigmoid), enabling the model to capture complex, nonlinear interactions among geometric, material, and reinforcement features. Model parameters are learned by backpropagation with gradient-based optimization (e.g., Adam) to minimize a regression loss (*RMSE*/*MSE*). Overfitting is controlled via early stopping, L2 regularization, and, if needed, dropout; inputs are z-score standardized to stabilize training. Network depth and width were selected to balance capacity and generalization, with hyperparameters tuned by repeated resampling. The MLP function is defined by Eq. (3):

$$\hat{y} = W_2\phi\left(W_1x + b_1\right) + b_2 \tag{3}$$

where x: input vector; $\phi(.)$: hidden-layer activation function; W_1, b_1: input to hidden weights and bias; and W_2, b_2: hidden to output weights and bias term.

XGBoost is an advanced implementation of the gradient-boosted decision tree (GBDT) algorithm, which incorporates the second-order Taylor expansion of the loss function to improve model performance. It has become highly effective and widely used, particularly for large datasets. It iteratively reduces the prediction error by fitting the residuals from the earlier round of estimations. Finally, the predicted values of individual trees are combined to obtain the final score [60, 61]. The objective function is presented in Eq. (4):

$$J(\theta) = \sum_i l\left(\hat{y}_i, y_i\right) + \sum_i R\left(g_k\right), \quad \hat{y}_i = \sum_k Rgk\left(x_i\right) \tag{4}$$

where l (.) is the loss measure and discrepancy between prediction $\hat{y}_i$ and experimental value y_i. R(.) is a regularization term that controls model complexity to mitigate overfitting; g_k donates the k-th tree (base learner); and θ collects the model parameters. Overall, XGBoost is the powerful tree-based algorithm [62].

6.2 *Practical Solution Development*

As mentioned earlier, PS was derived from simplified type of MLP [63]. This architecture is designed to capture the nonlinear relationship between input features and the EVDR of RC coupling beams. In PS, input features are connected to a single neuron in the hidden layer through weighted connections. The weights assigned to each input during the learning process create a summing junction, and a bias term is used to shift the weighted sum, while the neuron integrates with an activation function [64]. Notably, PS internally maps the original range of x to the range [−1,1], with the corresponding minimum and maximum values provided in Table 1. The structure of the PS is outlined as follows:

The neuron calculates the weighted sum from the input array $x = \left[x_1, x_2, x_3, \ldots, x_{14}\right]$ in the hidden layer through the following process:

$$z = \sum_{14}^{n=1} w_{i,HL}x_i + b_{HL} \tag{5}$$

where z indicates the weighted sum in the hidden layer (HL), w_i are the weights assigned to each input feature x_i, and b_{HL} is the bias term added to shift the weighted sum. To introduce nonlinearity, the weighted sum passes through an activation function:

$$h = \sigma(z) = \frac{1}{1 + e^{-z}} \tag{6}$$

where h is the result from hidden layer, and $\sigma(z)$ is the activation function applied to the weighted sum z. The output layer takes results from the hidden layer, applies the weight and bias of the output layer, and produces the final output.

$$\hat{y} = w_{OL} h + b_{OL} \tag{7}$$

The output layer takes the hidden layer result h, applies its own weights w_{OL} and bias b_{OL}, and computes the final output $\hat{y}_{PS}$.

To develop the PS, the dataset was randomly split into training, validation, and testing sets, 70%, 15%, and 15%, respectively. The training and testing sets were then used to train and test the solution network, enabling the calculation of the weights and biases for both the hidden and output layers in the PS. Using the trained network weights and biases, the practical approximation of EVDR can be mathematically expressed as shown in Eq. (5):

$$\begin{aligned} z = & -0.718(l^*) + 1.048(h^*) + 0.243(b^*) \\ & +0.880(AR^*) + 0.232(f_c'^*) + 0.515(f_{yl}^*) \\ & +0.068(f_{yv}^*) - 0.421(\rho_l^*) - 0.345(\rho_v^*) + 0.038(s) \\ & +0.010(\rho_d^*) - 1.090(f_{yd}^*) + 0.077(\alpha^*) + 2.446(\theta^*) + 1.236 \end{aligned} \tag{8}$$

$$h = \frac{1}{1 + e^{-z}}; \quad [\text{Activation fucntion}] \tag{9}$$

$$EVDR = 1.476(h) - 0.921 \quad \in [-1,1] \tag{10}$$

Each term in Eq. (8) represents the scaled input feature multiplied by the corresponding weights, along with the bias term from the hidden layer. Equation (9) was calculated using the logsig activation function, though various activation functions were tested. However, the logsig function provided stable and accurate results. Using the trained weights and biases from the network's output layer, Eq. (10) estimates the final EVDR, which is in the range of [−1,1]. To obtain the actual predicted EVDR, $EVDR_{PS}$ needs to be de-normalized using the minimum and maximum values of the original EVDR by using Eq. (11):

$$x^* = 2 * \frac{(x - x_{\min})}{x_{\max} - x_{\min}} - 1 \tag{11}$$

where x^* represents the scaled value of x. x_{min} and x_{max} represent the min and max values, respectively. To make the process smooth and more understandable, a flowchart is provided in Fig. 5 to estimate the EVDR of RC coupling beams. The

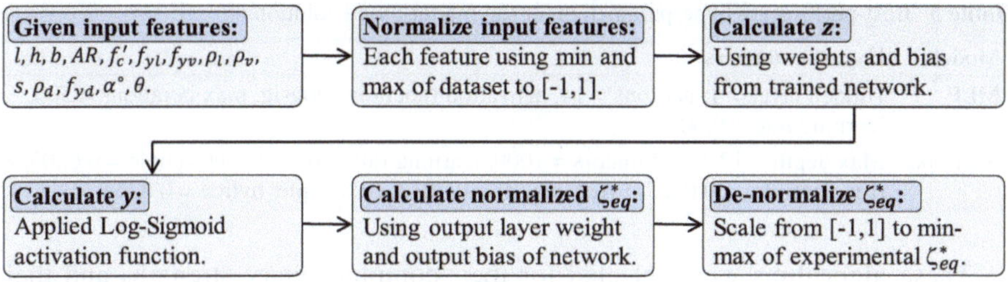

Fig. 5 Flowchart of the proposed PS for estimating EVDR

Table 2 Tuned hyperparameters of the PS solution

Solution	Hyperparameters
PS	Hidden layer = 1, neurons = 1, activation function = logsig, trainParam.epochs = 500, training objective function = Trainbr

flowchart effectively presents the essential steps, facilitating the straightforward implementation of PS. Optimal hyperparameters of the adopted PS are given in Table 2.

6.3 Solution Implementation and Performance Assessment

In ML-based models, selection and optimization of hyperparameters are very critical, as suitable values can mitigate the issue of overfitting and underfitting. Properly tuned hyperparameters increase the model capacity and generalization, so their selection directly controls the bias variance trade-off. A feed-forward neural network with an enhanced architecture was trained using Bayesian regularization to mitigate overfitting. Multiple training algorithms and activation functions were evaluated across several random initializations, with validation monitoring guiding selection of a robust, well-balanced configuration. Mean values of R^2 and MSE were recorded over 500 random trials using just one neuron in a hidden layer, then the number of neurons was gradually increased, and performance was checked at each step to enhance the model performance. Once the neuron count reached 14, performance showed stability, since there was no further improvement with adding more neurons. A similar procedure was conducted with decreasing the neurons. This way, the optimal number of neurons was found [65]. For the gradient-boosted decision tree model XGBoost, hyperparameters were tuned with Optuna via Bayesian optimization, which leverages prior evaluations to efficiently search the parameter space and identify near-optimal settings. The selected configurations were then fixed, and both models were retrained (using the combined training–validation data) to confirm predictive performance on the held-out test set. The chosen parameter settings for both algorithms are reported in Table 3.

Table 3 Best-performing hyperparameters for the model-based solutions

Models	Hyperparameters
MLP	Hidden layer = 1, neurons = 14, activation function = logsig, max iterations = 500, learning rate = 0.001
XGBoost	Max depth = 12, n_estimators = 1000, learning rate = 0.0116, subsample = 0.6205, λ (reg_lambda) = 0.0603, τ (reg_alpha) = 0.0012, colsample_bytree = 0.8264

These algorithms were selected for their complementary strengths and their ability to capture complex relationships in the data, yielding accurate and robust estimates. For both models, the dataset was randomly split into training, validation, and test subsets in 70%/15%/15%, respectively. Each model was evaluated across 500 randomized trials to identify the best performer and assess robustness. Model selection was based on the lowest MSE. The best configuration for each algorithm was then refitted on the combined training and validation data and evaluated on the untouched test set to provide an unbiased estimate of generalization.

6.4 Prediction Performance of Solutions

This section evaluates the predictive performance of the developed solutions using experimental vs. predicted EVDR. Figure 6 presents scatterplots of the test set where the x-axis shows experimental and the y-axis shows predicted EVDR. The ideal line 1:1 is shown together with ±10% and ±20% error bands to visualize scatter dispersion. Points concentrated along the 1:1 line, especially within the ±10% error band, show solid agreement.

The PS solution shows a modest dispersion with some data points exceeding the ±10% and ±20% error band limits. This behavior is expected in the PS given the intentionally developed minimalistic architecture—a single neuron in a hidden layer chosen to produce a compact and practical-use equation. Even so, the highest point density remains close to the 1:1 line and largely within the error band limits, indicating an acceptable central tendency and usable accuracy. A slightly wider spread is observed at higher EVDR values, while the center of the distribution remains close to the 1:1 line. Overall, PS provides an interpretable baseline that captures the main trend reliably, trading some precision for transparency and ease of use. The MLP models the nonlinear EVDR relationship effectively and yields stable responses. It is well calibrated with low scatter dispersion and has $R^2 = 0.97$. Most predictions cluster tightly along the ideal line and fall within the ±10% band, indicating close agreement with the experimental values. Residuals remain small across the range, with only minor deviation at the upper end of EVDR. Among the single-learner models evaluated here, the MLP provides the strongest overall performance. XGBoost showed the strongest agreement with experiments among all predictive approaches, yielding the highest R^2 value of 0.98. In Fig. 6, the predicted EVDR values lie close to the 1:1 diagonal

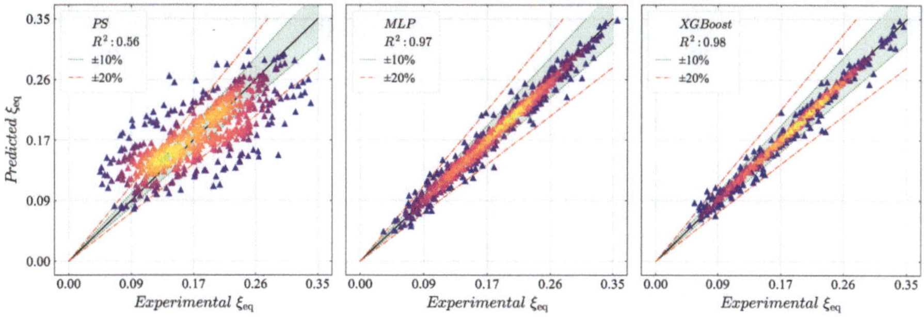

Fig. 6 Experimental vs. predicted EVDR across solutions

across the full experimental range, indicating that the model captures both the overall scale and the local variability of the response. The points are densely and narrowly clustered—especially between approximately 0.08 and 0.30 EVDR, where most observations lie—and the majority of predictions fall within the 10% and 20% error bands. Overall, the XGBoost model provides a high-fidelity fit with strong global calibration.

6.5 Solution Bias Evaluation

To further explore and assess the accuracy and error analysis of the developed solutions, bias evaluation was conducted. Figure 7a presents a 3D bar plot comparison of five metrics for each solution *ME*, *MAE*, *RMSE*, dispersion of error σ_{ME}, and OBG. This comparison delivered a detailed overview of the performances and revealed significant differences in prediction accuracy and bias, illustrating how each model performs in estimating the EVDR of the DRCBs. Among the developed solutions, XGBoost shows the superior accuracy and minimal bias as $ME = 0.001$ and residual dispersion of $\sigma_{ME} = 0.008$, almost zero. Even so, *ME* is close to zero indicating that there is virtually no systematic over- and underprediction, while σ_{ME} shows that individual errors are tightly clustered, reflecting consistency in case-by-case performance. Likewise, the error magnitude is very small: $RMSE = 0.008$ and $MAE = 0.005$. The *MAE* indicates that, on average, XGBoost EVDR predictions deviate from the measurements by only about 0.005 in absoluter term, while RMSE—being slightly larger—shows that occasional larger error is present but still minor. The fact that *RMSE* is close to *MAE* suggests no heavy-tailored outlier and generally tight residual distribution. Together, these values also confirm higher accuracy and stable per-case performance across database. An *OBG* of 0.007 indicates excellent overall balance of the residuals: the prediction errors are not only minor on average but also well centered with minimal skewed across the EVDR range.

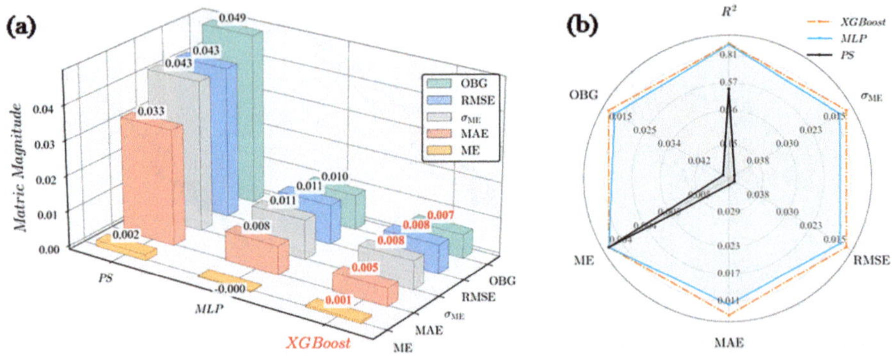

Fig. 7 Comparative evaluation of the developed solutions: (**a**) design suitability assessment and (**b**) radar chart of performance

The MLP solution exhibits reasonable accuracy with MAE = 0.008 and $RMSE$ 0.011, indicating negligible systematic bias in the predictive values. However, the error dispersion remains noticeable. σ_{ME} is 0.011, and OBG = 0.010 is slightly higher than that of XGBoost but still shows acceptable predictive performance. The PS solution shows ME = 0.002, MAE = 0.033, and $RMSE$ = 0.043. These values reflect its design intent of interpretability and ease of implementation with transparent feature effects, trading off predictive tightness for simplicity and quick hand calculations to estimate the EVDR. PS remains solid solution when tighter EVDR estimates are required; PS calculation steps in Fig. 5 clearly demonstrate the ease of implementation.

Overall, across both solutions, XGBoost emerges as a balanced and reliable predictive model of EVDR, combining minimal bias with consistently tight clustering around the experimental values. The model is well calibrated across the full response range, successfully capturing the complex interactions between input and output features. Its accuracy and robustness make it well suited for design evaluation and decision making, delivering dependable estimates while maintaining robustness. Figure 7b shows a radar chart comparison of the developed solutions, where larger radial extent indicates stronger fit and shorter inward spokes indicate smaller errors and bias. The XGBoost contour encloses the largest area with the most compact spokes, reflecting a highly balanced trade-off between accuracy and calibration across the evaluation dimensions. The MLP shows an intermediate profile, while the PS solution traces a smaller envelope. Overall, the radar chart visual verifies that XGBoost is the most reliable and well-suited solution.

6.6 Evaluation Metrics

To assess and compare performance and to quantify variations in the predictive behavior of the developed solutions, four widely used evaluation metrics were utilized: the coefficient of determination (R^2), root mean square error ($RMSE$), mean error (*ME*), and overall balanced grade (*OBG*). In general, these metrics capture goodness of fit (R^2), squared error magnitude (*RMSE*), signed bias (*ME*), and an overall balanced grade (*OBG*) that balances accuracy with fit value. All evaluation metrics were computed on the test set for each solution to ensure a fair assessment. The mathematical expressions are the following:

$$R^2 = 1 - \sum_{i=1}^{n} \left(y_i^{'} - y_i\right)^2 / \sum_{i=1}^{n} \left(y_i - \overline{y}_i\right)^2 \tag{12}$$

$$RMSE = \sqrt{\frac{\sum_{i=1}^{n} \left(y_i^{'} - y_i\right)^2}{n}} \tag{13}$$

$$ME = \frac{1}{n} \sum_{i=1}^{n} \left(y_i^{'} - y_i\right) \tag{14}$$

$$MAE = \frac{1}{n} \sum_{i=1}^{n} \mid y_i^{'} - y_i \mid \tag{15}$$

$$OBG = \sum \frac{MAE + RMSE}{1 + R^2} \tag{16}$$

where y_i is the actual experimental value, $y_i^{'}$ is the predicted value, $\overline{y}_i$ is the mean of the actual values, and n is the total number of data samples.

7 Solution Validation

This section introduces the predictive capability of the trained models on unseen data. Following hyperparameter tuning, optimization, and accuracy verification, the finalized models were saved. Model robustness and reliability under realistic conditions were then assessed via blind validation on two specimens reported in the literature: the DRCB specimen CB-70A from Yang et al. [47] and CB-X2 from Tianxiang Pi [51]. Figure 8 presents the blind validation results.

Both specimens were held out entirely from the training process. The trained MLP and XGBoost models were then used to predict EVDR as a function of θ, and the predictions were compared with the corresponding experimental measurements. Both models reproduced the trends with high out-of-sample agreement: for CB-70A, $R^2 = 0.96$ (MLP) and $R^2 = 0.94$ (XGBoost) with small mean offsets $y_{exp} - y_{pred} = +0.004$ and +0.012, respectively; for CB-X2, $R^2 = 0.96$ (MLP) and $R^2 = 0.95$ (XGBoost) with offsets of −0.000 and −0.008. The close overlap of predicted and experimental curves across the θ range indicates minimal bias and robust generalization. Because both models were trained on data spanning the full θ interval reported in Table 1 and their generalization was confirmed by blind validation, they are expected to provide reliable EVDR predictions for any experimental case whose θ lies within that range.

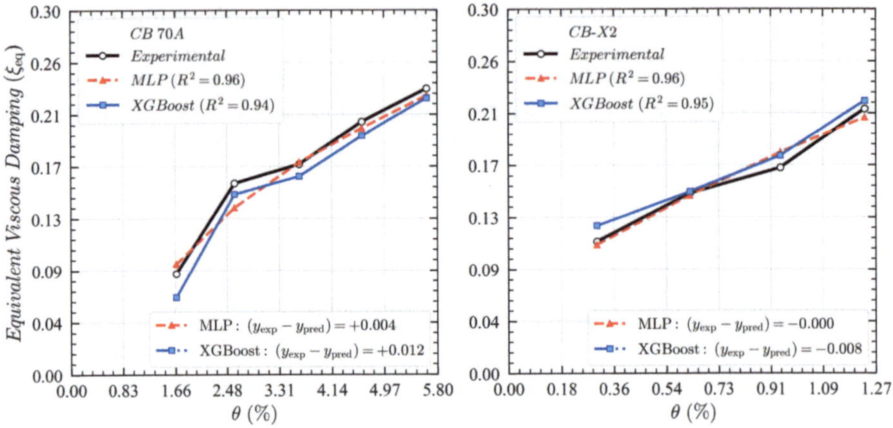

Fig. 8 Blind validation on CB-70A and CB-X2: predicted EVDR vs. θ from the trained model compared with experiments

The reported failure modes were flexural-shear for CB-70A and diagonal bar buckling for CB-X2. The developed solutions performed well for both mechanisms, indicating robustness across the intended DRCB failure types.

8 Conclusion

A peer-reviewed database of 125 tested DRCB specimens reported between 1980 and 2025 was compiled. Using this dataset, two machine learning solutions were developed to estimate the EVDR of coupling beams: a single-model MLP and an ensemble XGBoost. In addition, a PS was distilled from a simplified MLP formulation to yield an interpretable, closed-form equation suitable for practical EVDR estimation. Sensitivity analysis of the input features was performed to identify the most influential predictors, and blind validation was carried out to demonstrate the generalization capability of the proposed solutions. By comparison of experimental and predicted outcomes, the following conclusions are attained:

- Feature sensitivity analysis conducted based on feature weights indicates that the material strength features $\left(f_c^{'} f_{yd}, and\ f_{yl}\right)$, reinforcement ratios ($p_d$, p_v, *and* p_l), and geometric aspect ratio (*AR*) are the most influential features for predicting the EVDR of DRCBs.
- Comparison analysis highlights that both models perform well; however, the XGBoost attains the superior accuracy, with $R^2 = 0.98$, $ME = 0.001$, $\sigma_{ME} = 0.008$, $RMSE = 0.008$, $MAE = 0.005$, and OBG of 0.007. These metrics indicate almost no bias and a tightly clustered residual dispersal. Considering both predictive accuracy and blind validation generalization, XGBoost outper-

forms the other solutions, and it is regarded as an unbiased, highly reliable, and balanced model.

- The PS was derived from a simple and interpretable architecture of MLP, yielding an explicit mathematical equation for estimating the EVDR without requiring any complex programming knowledge. In performance evaluation, PS achieved $R^2 = 0.56$, $RMSE = 0.043$, and $ME = 0.002$. Users can compute EVDR directly by using the input features and applying the given activation function with the provided trained weights and biases of the network, enabling easy implementation.

The blind validation of the developed solutions demonstrated that the proposed solutions are not only accurate but also highly reliable, with predicted EVDR closely matching experimental values. For CB-70A (flexural-shear failure), the results yielded $R^2 = 0.96$ (MLP) and 0.94 (XGBoost); for CB-X2 (diagonal bar buckling), $R^2 = 0.96$ (MLP) and 0.95 (XGBoost). These outcomes indicate strong generalization and support the applicability of the proposed approaches to new data within the studied θ range.

Both solutions in this study provide high accuracy and strong generalization to unseen data. Further work is encouraged to improve the interpretability of the models' black box nature. Future efforts could focus on reliability analysis of the ML models, including uncertainty quantification and calibration. Researchers are encouraged to expand the database: conventional and crossover reinforcement can be included to analyze and investigate the response toward a more universal solution.

Acknowledgements This research was supported by the National Natural Science Foundation of China (NSFC) under grant numbers U2139209 and 52378528. The author, Bilal Younis, also gratefully acknowledges the China Scholarship Council (CSC) for providing financial support for his postgraduate studies.

AI Use Disclosure Statement During manuscript preparation, the authors used ChatGPT to improve clarity and readability. All content was subsequently reviewed and edited by the authors, who take full responsibility for the accuracy, integrity, and final version of the publication.

Data Availability The database underlying this study is available from the corresponding author upon reasonable request.

References

1. Shiu KN, Takayanagi T, Corley WG (1984) Seismic behavior of coupled wall systems. J Struct Eng 110:1051–1066
2. Lu X, Chen C, Jiang H, Wang S (2018) Shaking table tests and numerical analyses of an RC coupled wall structure with replaceable coupling beams. Earthquake Eng Struct Dyn 47:1882–1904

3. Gonzalez E (2001) Seismic response of diagonally reinforced slender coupling beams
4. Galano L, Vignoli A (2000) Seismic behavior of short coupling beams with different reinforcement layouts. Struct J 97:876–885
5. Breña SF, Ihtiyar O (2011) Performance of conventionally reinforced coupling beams subjected to cyclic loading. J Struct Eng 137:665–676
6. Paulay T, Binney J (1974) Diagonally reinforced coupling beams of shear walls. Special Publication 42:579–598
7. Hindi RA, Hassan MA (2004) Shear capacity of diagonally reinforced coupling beams. Eng Struct 26:1437–1446
8. Han SW, Lee CS, Shin M, Lee K (2015) Cyclic performance of precast coupling beams with bundled diagonal reinforcement. Eng Struct 93:142–151
9. Mai-Quang K, Han SW, Shin M, Lee K (2017) Reduction of reinforcement congestion in slender coupling beam using bundled diagonal bars. Mag Concr Res 69:1157–1169
10. Jafari A, Beheshti M, Shahmansouri AA, Bengar HA (2024) Cyclic response and damage status of coupled and hybrid-coupled shear walls. Structures 2024:106010
11. Jafari A, Beheshti M, Shahmansouri AA, Bengar HA (2022) Plastic hinge length for coupled and hybrid-coupled shear walls. Steel Compos Struct:367–383
12. Jobcobsen L (1930) Steady forced vibration as influenced by damping. Trans ASME 52:169–181
13. Rosenblueth E, Herrera I (1964) On a kind of hysteretic damping. J Eng Mech Div 90:37–48
14. Iwan W (1980) Estimating inelastic response spectra from elastic spectra. Earthquake Eng Struct Dyn. 8:375–388
15. Miranda E, Ruiz-García J (2002) Evaluation of approximate methods to estimate maximum inelastic displacement demands. Earthquake Eng Struct Dyn 31:539–560
16. Howard B (2017) Seismic response of diagonally reinforced coupling beams with varied hoop spacings. McGill University (Canada)
17. Dugas DG (2002) Seismic response of diagonally reinforced coupling beams with headed bars
18. Amiri A, Atkinson J, Tobber L (2025) Evaluating the influence of detailing on the seismic performance of diagonally reinforced coupling beams. Eng Struct 340:120655
19. Zhao Y, Tian J, Zhou W, Liu G, Lu J (2025) Prediction of shear capacity of PRC coupling beams using machine learning methods. Structures 2025:108783
20. Hu X-W, Ding R, Zhang Z-Y, Fan J-S (2023) A unified section-based restoring force model for reinforced concrete coupling beams based on the XGBoost model: establishment, validation and application. J Build Eng 73:106666
21. Yaghoubi ST, Deger ZT, Taskin G, Sutcu F (2023) Machine learning-based predictive models for equivalent damping ratio of RC shear walls. Bull Earthq Eng 21:293–318
22. Li Y, Liu Y, Yu H, Ma K, Zhang X, Ji W et al (2023) Multi-objective optimization design of coupled wall structure with hybrid coupling beams using hybrid machine learning algorithms. J Build Eng 78:107745
23. Jafari A, Shahmansouri AA, Taslimi A, Naser M, Zhou Y (2025) Shear capacity of FRP-RC beams: XML modeling and reliability evaluation. Structures 2025:110150
24. Canbolat BA, Parra-Montesinos GJ, Wight JK (2005) Experimental study on seismic behavior of high-performance fiber-reinforced cement composite coupling beams. ACI Struct J 102:159
25. Shin M, Gwon S-W, Lee K, Han SW, Jo YW (2014) Effectiveness of high performance fiber-reinforced cement composites in slender coupling beams. Constr Build Mater 68:476–490
26. Shimazaki K (2004) De-bonded diagonally reinforced beam for good repairability. In: Conference 13th annual institute. Earthquake engineering of Vancouver (BC), Canada
27. Lim E, Hwang S-J, Cheng C-H, Lin P-Y (2016) Cyclic tests of reinforced concrete coupling beam with intermediate span-depth ratio. ACI Struct J 113
28. Lim E, Hwang S-J, Wang T-W, Chang Y-H (2016) An investigation on the seismic behavior of deep reinforced concrete coupling beams. ACI Struct J 113:217
29. Kwan A, Zhao Z-Z (2002) Cyclic behaviour of deep reinforced concrete coupling beams. Proc Inst Civ Eng Struct Build 152:283–293

30. Ameen S, Lequesne RD, Lepage A. Diagonally reinforced concrete coupling beams with grade 120 (830) high-strength steel bars. American Concrete Institute; 2020
31. Li Y, Zheng S, Dong L, Wang D, Sang Z (2025) Experimental investigation of the seismic performance of corroded reinforced concrete coupling beams. Eng Struct 322:119065
32. Park W-S, Yun H-D (2011) Seismic performance of pseudo strain-hardening cementitious composite coupling beams with different reinforcement details. Compos Part B 42:1427–1445
33. Fortney PJ, Rassati GA, Shahrooz BM (2008) Investigation on effect of transverse reinforcement on performance of diagonally reinforced coupling beams. ACI Struct J 105:781
34. Cheng M-Y, Gitomarsono J, Zeng H-Y (2019) Cyclic test of diagonally reinforced concrete coupling beam with different shear demand. ACI Struct J 116:241–250
35. Shin DI, Haroon M, Yoon SH, Lee J-Y (2020) Effect of deviation angle between principal stress and diagonal reinforcement direction on seismic performance of RC coupling beams. Eng Struct 215:110618
36. Han SW, Kim SB, Kim T (2019) Effect of transverse reinforcement on the seismic behavior of diagonally reinforced concrete coupling beams. Eng Struct 196:109307
37. Sesli H, Husem M (2021) Experimental investigation on cyclic behavior of reinforced concrete coupling beams under quasi-static loading. Int J Civ Eng 19:381–400
38. Poudel A, Lequesne RD, Lepage A. (2018) Diagonally reinforced concrete coupling beams: effects of axial restraint. SL Report 18 3
39. Saathoff DB (2022) Nonlinear modeling parameters for reinforced concrete coupling beams. Washington State University
40. Barney G, Shiu K, Rabbit B, Fiorato A, Russell H, Corley W (1980) Behavior of coupling beams under load reversals (RD068. 01B). Portland Cement Association, Skokie, IL
41. Ishikawa Y, Kimura H (1996) Experimental study on seismic behavior of R/C diagonally reinforced short beams. Proceedings of the 11th world conference on earthquake engineering, p 92
42. Kanakubo T, Fujisawa M, Sako N, Sonobe Y (1996) Ductility of short span RC beams
43. Naish D, Fry A, Klemencic R, Wallace J (2013) Reinforced concrete coupling beams—Part I: Testing. ACI Struct J 110:1057
44. Weber-Kamin A (2020) Reinforced concrete coupling beams with high-strength steel bars. University of Kansas
45. Park W-S, Kang TH-K, Kim S, Yun H-D (2020) Seismic performance of moderately short concrete coupling beams with various reinforcements. ACI Struct J 117
46. Abdullah S, Aswegan K, Klemencic R, Wallace J (2020) Experimental study of concrete coupling beams subjected to wind and seismic loading protocols. Report no UCLA SEERL 1:268
47. Yang C, Chen S-C, Yen C-H, Hung C-C (2022) Behaviour and detailing of coupling beams with high-strength materials. J Build Eng 47:103843
48. Setkit M (2012) Seismic behavior of slender coupling beams constructed with high-performance fiber-reinforced concrete. University of Michigan
49. Jang S-J, Jeong G-Y, Kim S-W, Yun H-D, Chun Y-S (2017) Effect of reinforcement layout on structural performance of reinforced concrete coupling beams with high-strength steel bar. J Korea Inst Struct Maintenance Inspect 21:95–102
50. Tassios TP, Moretti M, Bezas A (1996) On the behavior and ductility of reinforced concrete coupling beams of shear walls. Struct J 93:711–720
51. Pi T (2008) Experimental and design study on seismic performance of reinforced concrete shear wall coupling beams with small span-to-height ratio. PhD Thesis in Chinese
52. Jang S-J, Jeong G-Y, Yun H-D (2018) Use of steel fibers as transverse reinforcement in diagonally reinforced coupling beams with normal- and high-strength concrete. Constr Build Mater 187:1020–1030
53. Komiya Y, Nakatsuka T, Onal T, Naruse T, Ryujin H (2007) Experimental study on structural characteristics of diagonally reinforced short span beam. J Struct Constr Eng 5:173–180
54. Jobcobsen L (1960) Damping in composite structures. In: Proceedings of second world conference on earthquake engineering, Tokyo, pp 1029–1044

55. Topaloglu B, Kaya GT, Sutcu F, Deger ZT (2022) Machine learning-based estimation of energy dissipation capacity of RC shear walls. Structures:343–352
56. Son D-H, Choi J-Y, Song S, Choi C-S, Bae B-I (2024) Inelastic behavior of diagonally reinforced coupling beams confined with steel fibers. Structures 65:106733
57. Han SW, Kang JW, Lee CS (2018) Cyclic behavior of diagonally reinforced slender HPFRCC coupling beams with reduced diagonal and transverse reinforcement. Compos Struct 206:550–562
58. Choi Y, Chao S-H (2020) Analysis and design of double-beam coupling beams. ACI Struct J 117:79–95
59. Ramchoun H, Ghanou Y, Ettaouil M, Janati Idrissi MA (2016) Multilayer perceptron: architecture optimization and training
60. Ester M, Kriegel H, Xu X (2022) XGBoost: a scalable tree boosting system. In: Proceedings of the 22nd ACM SIGKDD international conference on knowledge discovery and data mining. GEOGRAPHICAL ANALYSIS, p 785
61. Chen T, Guestrin C (2016) Xgboost: a scalable tree boosting system. In: Proceedings of the 22nd ACM SIGKDD international conference on knowledge discovery and data mining, pp 785–794
62. Feng D-C, Wang W-J, Mangalathu S, Hu G, Wu T (2021) Implementing ensemble learning methods to predict the shear strength of RC deep beams with/without web reinforcements. Eng Struct 235:111979
63. Jafari A, Mollaali M, Ma L, Shahmansouri AA, Zhou Y, Dugnani R (2025) Brittle fracture strength prediction via XML with reliability considerations. Eng Fract Mech 328:111555
64. Jafari A, Mirrashid M, Hoult RD, Zhou Y (2025) Plastic hinge length of RC shear walls: practical approximation via machine learning and probabilistic assessment. Eng Fail Anal 169:109179
65. Shahmansouri AA, Jafari A, Akbarzadeh Bengar H, Zhou Y (2025) Steel rebar bond-slip behavior in CFRP-strengthened low-strength concrete: ML-driven solution and reliability analysis. Constr Build Mater 490:142302

MIX
Papier aus verantwortungsvollen Quellen
Paper from responsible sources
FSC® C105338

If you have any concerns about our products,
you can contact us on
ProductSafety@springernature.com

In case Publisher is established outside the EU,
the EU authorized representative is:
Springer Nature Customer Service Center GmbH
Europaplatz 3, 69115 Heidelberg, Germany

Printed by Libri Plureos GmbH
in Hamburg, Germany